Oliver Rivington Willis

Report of the Flora of Westchester County

Oliver Rivington Willis

Report of the Flora of Westchester County

ISBN/EAN: 9783337268695

Printed in Europe, USA, Canada, Australia, Japan

Cover: Foto ©berggeist007 / pixelio.de

More available books at **www.hansebooks.com**

REPORT

OF THE

Flora of Westchester County.

PREPARED FOR

BOLTON'S HISTORY OF THE COUNTY.

BY

OLIVER R. WILLIS, A. M., PH. D.,

Author of "CATALOGUS PLANTANUM IN NOVA CAESAREA REPERTARUM.

INTRODUCTION.

T HE following report of the Flora of Westchester County has been
prepared at the request of the Rev. Robert Bolton, for publication
in the revised edition of his history of the county.

It was with great reluctance that the author undertook the work;
though he has, during the last seventeen years, botanized in the region,
and has corresponded with and examined the collections of botanists
who have visited parts of the territory not examined by himself; yet he
feels sure that there are plants growing in the limits which have not
been noticed by botanists, or, at least, have not been reported

He set out, however, with the determination not to include any in
the list except such as he himself, or some one in whom he had full
confidence, had seen growing in the county. As no one has examined

771

the region with the intention of publishing a report, until within the last two years, it is reasonable to infer that many plants will yet be found that are not included in this report. The author has corresponded and held conferences with not only the botanists of the county, but with those of New York city.

The botanists of New York city are active, industrious collectors, and have absolutely exhausted most of the localities within walking distance of the city; especially such portions of New Jersey, Staten Island and parts of Long Island. The Harlem River, however, seems to have been, to a great extent, the limit to their excursions in this direction; hence, this county is less known to New York botanists than any other region within the same distance of the city. The collecting of information has, therefore, been more formidable than suspected.

The State of New York has a remarkably large Flora; but it has a territory extending about 350 miles from north to south, and nearly the same distance from east to west—giving it a wide range of climate and temperature. It has plains along the coast elevated just a little above the sea level, and mountains on the eastern border rising 6,000 feet above the ocean; it possesses every variety of soil, from the sands of the Saranac region to the alluvial plains of the western slope and the rich bottom lands of the head waters of the Susquehanna, and the valleys of the Mohawk and the Upper Hudson.

Dr. Torrey, in his report of the Flora of the State, stated that the number of flowering plants would reach 1,450; and the Ferns and Lycopodiaceæ, sixty.

There will be great reason, then, for wonder, when it is found that the number of plants growing without cultivation in Westchester County— a territory about fifty miles in extent from north to south, and whose average width from east to west is not half so much—is more than a thousand.

There are 1,142 flowering plants enumerated in this catalogue, and forty-six ferns and their allies.

The formation of the county is chiefly Gneisic and Limestone rocks. Limestone is sprinkled throughout, but especially along the middle, from north to south. The southern half is divided by two parallel valleys, which trend north and south—dipping towards the south—and about three miles apart, separated by a ridge of hills. The valley on the Eastern side of the ridge is drained by the Bronx River, and the other (in part) by Saw-Mill River continued by Tibbitt's Brook. This limited territory has no difference of climate and temperature.

The soil is made up of abrasions and disintegrations of Gneisic and Limestone rocks and sand, sparingly mixed with clay; forming what agriculturalists denominate "Light Loam"—a soil especially adapted to the growth of cereals—yielding abundant crops to generous cultivation. But in such narrow limits, we cannot have a wide range in the variety of soil; in fact, we have very little more than such variety as is produced by hill and vale, or wet, damp, hilly and rocky—which is not so much a variety of soil as a difference in the state of the same soil.

We necessarily infer that such conditions are not favorable to a prolific Flora.

There are other features, however, that must not be overlooked, that ought to give rise to variety in the plants growing in the county.

The territory is washed on one side by the Hudson—a long river flowing through nearly the whole length of the eastern side of the State. On the other side, the coast of the Sound gives it the sea beach and the salt marsh Three railways extend through it from north to south; one of which is part of the great highway between the Atlantic and the Pacific oceans; one of the others, by its connections, reaches the shores of the Gulf of St. Lawrence, and the third extends through the middle and reaches Canada. We should, therefore, expect that the Flora would be enriched and greatly enlarged by seeds brought and sown along the lines of these great railways and floated to the shores by the waters of the Hudson and Long Island Sound. In this, however, we are disappointed; for it is remarkable that a very small number of plants are growing in the county, that have been introduced by the means above mentioned.

Of the plants herein enumerated, eighty-eight have been introduced from Europe; fifty-seven of this number are growing without cultivation and freely propagating themselves, and are said to be naturalized; the remainder, thirty-one in number, are growing without cultivation, but are not fully established.

As already stated, there has been detected in the county 1,142 flowering plants and forty-six ferns and their allies. Of the flowering plants, fifty are first-class trees, reaching a height of thirty feet and upwards; thirty-four are second-class trees which attain a height of fifteen to thirty feet; and sixty-nine are shrubs, all of which are growing without cultivation, a very few of which have been introduced.

The author would again call attention to what was stated in the beginning of this introduction, viz: all the plants named in this report are known to be growing within these limits without cultivation.

By including hardy trees and shrubs which have been introduced, and are growing in planted grounds, the list would be very much enlarged.

EXPLANATIONS.

The arrangement of this catalogne corresponds with that of the latest edition of *Gray's Manual of the Northern United States*, and the orders are numbered to agree with the numbers of the same orders in the *Manual*.

The words "Nat. Eu." mean the same as they do in the *Manual*, viz: that the plants after whose names they appear, have been introduced from Europe and are growing and propagating themselves freely without cultivation, and are fully established. "Adv. Eu." indicate that plants after whose names they are written, are from Europe; that they are growing without cultivation, but are not propagating themselves with such freedom and constancy as to be considered fully established

AIDS AND SOURCES OF INFORMATION.

In the collecting of the material for this report, I have received valuable information and assistance from the Botanists of the county and New York city. One of the most valuable sources of information has been the *Bulletin* of the Torrey Botanical Club.

But my thanks are especially due to Prof. Alphonso Wood, Ph.D., of West Farms, Prof. W. H. Leggett, editor of the *Bulletin* of the Torrey Botanical Club, and Curator P. V. Le Roy. To Miss P. A. Mecabe of Scarsdale, Mr. Chas. R. Hexamer of New Castle, Dr. G. J. Fisher of Sing Sing, and E. P. Bicknell of Yonkers.

CATALOGUE OF PLANTS.

SERIES I.

PHŒNOGAMOUS OR FLOWERING PLANTS.

CLASS I.—DICOTYLEDONOUS OR EXOGENOUS PLANTS.

Order 1.—RANUNCULACEÆ. (Crowfoot Family.)

Clematis, L.. (Virgin's Bower. Traveler's Joy.)
    C. verticillaris, DC. Rare. (Dr. Mead.)
    C. Virginiana, L.. Common in damp thickets. Bears transplanting **well.**

Anemone, L.. (Wind Flower.)
    A. cylindrica, Gray. Edges of woods. Rare.
    A. Virginiana, L.. Damp open woods. Rare.
    A. Pennsylvanica, L.. Not common. (Le Roy.)
    A. nemorosa, L. Common.

Hepatica, Dill. (Liverleaf.)
    H. triloba, Chaix. Woods common.
    H. acutiloba, DC. Not common.

Thalictrum, Tourn. (Meadow Rue.)
    T. anemonoides, Mx. Woods common.
    T. dioicum, L. Frequent in damp woods.
    T. purpurascens, L. Not common.
        Var. ceriferum. Austin. Wnite Plains ; also Riverdale. (T. C. B.)
    T. Cornuti, L.. Common in meadows.

Ranunculus, L. (Crowfoot, Buttercup,)
    R. aquatilis, L.. White Plains, slow brooks. Not common in this **County.**
    R. alismæfolius, Geyer. White Plains. Very wet places, not common.
    R. abortivus, L.. Frequent.
        Var. micranthus, Gray. Not rare.
    R. sceleratus, L. Rather rare.
    R. recurvatus, Poir. Frequent.
    R. Pennsylvanicus, L. Not common.

Ranunculus, L.   (Crowfoot Buttercup.)—*Continued.*

R. fascicularis, Muhl. Frequeut.

B. repens, L.   Common in damp places.

R. bulbosus, L.   Rare, (Nat. Eu.)

R. acris, L.   (Buttercups,)   Common.   (Nat. **Eu.**)

Caltha, L.   (Cowslips, Marsh Marigola.)

C. palustris, L.   Frequent, in wet places.

Trollius, L.   (Globe flower.)

T. laxus, Salisb.   Frequent.

Coptis, Salisb.   (Gold thread.)

C. trifolia, Salisb.   Frequent.

Aquilegia, Tourn.   (Columbine.)

A. Canadensis, L.   Common among rocks.   Bears transplanting.

Actæa, L.   (Baneberry.)

A. spicata, L.

Var, rubra, Mx.   Frequent, in shady woods.

A. alba, Bigel.   (White Baneberry.)   White Plains.

Cimicifuga, L.   Bugbane.

C. racemosa, Ell.   (Black snake root.)   Frequent in copses and woods.

Order 2.—MAGNOLIACEÆ. (Magnolia Family.)

Liriodendron, L.   (White Wood, Tulip tree.)

L. tulipifera. L.   Not rare.   This is a fine tree for ornamental purposes, and its wood is used much by cabinet makers, for drawers and linings.

Order 4.—MENISPERMACEÆ. (Moonseed.)

Menispermum, L.

M. Canadense, L.   Frequent.

Order 5.—BERBERIDACEÆ. (Barberry.)

Berberis, L.

B. vulgaris, L.   Not common.   About New Rochelle.   (Nat. **Eu.**)

Caulophyllum, Mx.

C. thalictroides, L.   Not common.   (Bicknell.)

Podophyllum, L.   (Mandrake, Love apple.)

P. peltatum, L.   Near Peekskill.   Not common.   (Le. Roy.)

Brassenia, Purgh.

B. peltata.   Mohegan Lake.   (Leggett.)

Order 6.—NYMPHÆACEÆ. (Water Lily.)

Nymphæa, Tourn.

N. odorata, Ait. (Sweet scented water lily.) Common in mill ponds and lakes.

Nuphar, Smith.

N. advena, Ait. Common in ponds.
N. luteum, Smith.
Var. pumilum, Gray. Near Wood Lawn. Not common.

Order 7 —SARACENIACEÆ. (Side-Saddle Flower.—Pitcher plant.)

Saracenia, Tourn.

S. purpurea, L. Bedford. (Hexamer.—Jas. Wood.)

Order 8.—PAPAVERACEÆ. (Poppy Family.)

Chelidonium, L. (Celandine.)

C. majus, L. Common about houses. (Nat. Eu.)

Sanguinaria. Dill. (Blood Root.)

S. Canadensis, L. Common about White Plains.

Order 9.—FUMARIACEÆ. (Fumitory Family.)

Dicentra, Bork.

D. cucullaria, DC. (Dutchman's Breeches.) Damp banks, not common.

Corydalis, Vent.

C. glauca, Pursh. Rocks about White Plains.

Fumaria, L. (Fumitory.)

F. officinalis, L. Riverdale. T. C. B. (E. P. Bicknell.) (Ad. Eu.)

Order 10.—CRUCIFERÆ. (Mustard Family.)

Nasturtium, R. Br.

N. officinale, R. Br. Banks and edges of Brooks. (Nat. Eu.)
N. sylvestre, R. Br. (Yellow cress.) Peekskill, (Leggett.) (Nat. Eu.)
N. Armoracia, Fries. Waysides escaped from cultivation. (Nat. Eu.)

Dentaria, L.

D. diphylla, L. About White Plains, not rare.
D. laciniata, Muhl. About White Plains.

Cardamine, L. (Bitter cress.)

C. rhomboidea, DC. Wet places.
C. pratensis, L. Frequent.
C. hirsuta, L. New Castle.
Var. sylvatica. (Bicknell.)

Arabis, L. (Rock Cress.)

A. lyrata, L. Rocky woods, not rare.
A. dentata, Torr & Gray. Frequent.
A. lævigata, DC. Common.
A. Canadensis, L. Frequent throughout.

Barbarea, R. Br. (Winter Cress.)

B. vulgaris, R. Br. Too common ; a troublesome weed.
B. præcox, R. Br. (Early Winter Cress.) (Scurvy Grass,) escaped from cultivation—Riverdale, (Bicknell.)

Sysymbrium, L. (Hedge Mustard.)

S. officinale, Scop. About dwellings. (Nat. Eu.)
S. Thaliana, Gaud. About dwellings. (Nat. Eu.)
S. Alliaria, Brown. Kings Bridge. (Bicknell—Adv. Eu.)

Hesperis, L.

H. matronalis, L. Introduced from Europe. (Bicknell.)

Brassica, Tourn. (Mustard,)

B. Sinapistrum. Boissier. Cultivated grounds. (Adv. Eu.)
B. alba, Gray. In cultivated grounds. (Adv. Eu.)
B. nigra, Gray. Cultivated grounds. (Adv. Eu.)
B. campestris, L. Escaped from cultivation. (Bicknell.)
B. oleracea, L. Escaped from cultivation. (Bicknell.)

Draba, L. (Whitlow Grass.)

D. Caroliniana, Walt. Not rare.
D. verna, L. Mott Haven, along the railroad.

Camelina, Crantz. (False Flax.)

C. sativa, Crantz. Cultivated grounds. (Adv. Eu.)

Capsella, Vent. (Shepherd's Purse.)

C. Bursa-pastoris, Mœnch. About dwellings. (Nat. Eu.)

Lepidium, L. (Pepper Grass.)

L. Virginicum, L. About dwellings ; used sometimes for salad.
L. campestre, L. Bicknell. (Nat. Eu.)

Raphanus, L. (Radish.)

R. Raphanistrum, L. Common, and troublesome weed. (Adv. Eu.)
R. sativus. L. Escaped from cultivation. (Bicknell.)

Order 11.—CAPPARIDACEÆ. (Caper Family.)

Polanisia, Raf.

P. graveolens, Raf. Peekskill, (Mead.)

Order 12.—RESIDACEÆ. (Mignonette Family.)

Reseda, L.

R. Luteola, L.   Roadsides.   (Adv. Eu.)

Order 14.—VIOLACEÆ. (Violet Family.)

Solea. DC.   (Green Violet.)

S. concolor, Ging.   Near Tarrytown, not common.   (Hall.)

Viola, L.   (Violet.)

V. rotundifolia, Mx.   Near Spuyten-Duyvil, and White Plains, sparingly throughout.

V. lanceolata, L.   Near Peekskill.   (Le Roy.)

V. primulæfolia, L,   Wet grounds throughout the County.

V. blanda, Willd.   Common with the last.

V. cucullata, Ait.   Common.

Forms.

"   **a.** striata.   Petals white and marked with purple lines.

"   **b.** palmata, Gray.   Leaves varying from cordate entire to palmate or pedate divided.

"   **c.** cordata, Gray.   Very broad cordate, sometimes reniform.

V. sagittata, Ait.   Frequent about White Plains.

Var. ovata, Nutt.   In dry grounds.

V. canina, L.   Damp grounds, common.

V. rostrata, Pursh.   New Castle, not common.

V. striata, Ait.   New Castle, not common.

V. Canadensis, L.   New Castle, not common.

V. pubescens, Ait.   Common throughout these limits.

Var. scabriuscula, Torr. & Gray.   Frequent.

V. tricolor, L.   Near Peekskill, escaped from cultivation.   (Le Roy—Adv. Eu.)

V. odorata, L.   Riverdale, escaped from gardens.   (Bicknell).

Order 14.—CISTACEÆ. (Rock-rose Family.)

Helianthemum, Tourn.

H. Canadense, Mx.   Road-sides.

Lechea, L.

L. major Mx.   Frequent.

L. minor Lam.   (Bicknell.)

L. racemulosa, Mx.

Order 15.—DROSERACEÆ. (Sundew Family.)

Drosera, L.   (Sundew.)

D. rotundifolia, L.   Scarsdale.   (Miss P. A. McCabe.)

Order 16.—HYPERICACEÆ. (St. John's-Wort Family.)

Hypericum, L.

    H. prolificum, L. Riverdale. (Bicknell.)
    H. ellipticum, Hook. New Castle.
    H. corymbosum, Muhl. Pasture. (Leggett.)
    H. perforatum, L. Common in fields. (Nat. Eu.)
    H. mutilum. Damp grounds, common.
    H. Canadense, L. Damp sandy grounds.
    H. Sarothra, Mx. Common in sandy fields.

Elodes, Adans. (Marsh, St. John's-wort.)

    E. Virginica, Nutt. Swamps.

Order 17.—ELATINACEÆ. (Water wort Family.)

Elatine, L.

    E. Americana. Arnott.

Order 18.—CARYOPHYLLACEÆ.

Dianthus, L. (Pink.)

    D. Armeria, L. New Castle* (Adv. Eu.)

Saponaria, L. (Soap-wort.)

    S. officinalis, L. (Bouncing Bet,) common. (Adv. Eu.)

Vaccaria, Medik. (Cow-Herb.)

    V. vulgaris, Host. Peekskill. (Le Roy.) (Adv. Eu.)

Silene, L.

    S. stellata, Ait. Frequent about White Plains.
    S. inflata, Smith. About White Plains. (Nat. Eu.)
    S. Pennsylvanica, Mx. White Plains.
    S. Armeria, L. Peekskill. (Le Roy.) Escaped from cultivation. (Adv. Eu.)
    S. antirrhina, L. Peekskill, (Le Roy.)
    S. noctiflora, L. Escaped. (Bicknell.)

Lychnis, Tourn. (Cockle.)

    L. vespertina, Sibth. (Riverdale—Bicknell.) (Adv. Eu.)
    L. Githago, Lam. In grain fields. (Adv. Eu.)

Arenaria, L.

    A. serpyllifolia, L. (Nat. Eu.)

Stellaria, L. (Chickweed.)

    S. media, Smith. Common about dwellings. (Nat. Eu.)
    S. longifolia, Muhl. Not rare.

Cerastium, L. (Mouse-ear—Chickweed.)

C. vulgatum, L.
C. viscosum. L. (Nat. Eu.)
C. arvense, L. Peekskill. (Le Roy.)

Sagina, L. (Pearl-wort.)

S. procumbens, L. Bronx River, near Williams bridge, (Rickard.)

Spergularia, Pers. (Sand-Spurrey.)

S. rubra, Pers.
Var. campestris, Gray. (Peekskill.—Le-Roy.)

Spergula, L. Sandy field.
S. arvensis, L.

Anychia Mx. (Forked Chickweed.)

A. dichotoma, Mx.

Scleranthus, L.

S. annuus. (Nat. Eu.)

Mollugo, L. (Indian Chickweed.)

M. verticillata, L. Cultivated grounds.

Order 19.—PORTULACACEÆ. (Purslane Family.)

Portulaca. Tourn. (Purslane.)

P. oleracea, L. Common in gardens and cultivated grounds. (Nat. Eu.)
P. grandiflora. Escaped from cultivated grounds.

Claytonia, L. (Spring Beauty.)

C. Virginica, L. About White Plains.
C. Caroliniana, Mx. About White Plains.

Order 20.—MALVACEÆ. (Mallow Family.)

Althæa, L. (Marsh-Mallow.)

A. officinalis, L. Coast. (Nat. Eu.)

Malva, L. (Mallow.)

M. rotundifolia, L. Common about dwellings. (Nat. Eu.)
M. sylvestris, L. (Adv. En.) Road-sides.
M. moschata, L. (Mush-Mallow.) Wood-lawn. (Bicknell—Adv. Eu.)

Abutilon, Tourn. (Indian Mallow.)

A. Avicennæ, Gærtn. About dwellings. (Adv. India.)

Hibiscus, L. (Rose-Mallow.)

H. Moscheutos, L. Near the coast.
H. Trionum, L. Escaped from cultivation. (Adv. Eu.) (Dr. Fisher.)

Order 21.—TILICAEŒ, (Linden Family,)

Tilia. L. (Basswood—Linden.)

T. Americana, L.

Order 23.—LINACEÆ. (Flax Family,)
Linum. L. (Flax.)

L. Virginianum, L.   Frequent in borders of woods.
L. striatum, Walt.   Riverdale.   (Bicknell.)

Order 24—GERANEACEÆ, (Geraneum Family,)

Geranium, L. (Cranesbill.)

G. maculatum, L.   Bears transplanting, woods, common
G. Carolinianum, L.   White Plains.
G. Robertianum, L.   Dry soil, waste grounds.

Floerkea, Wild. (False Mermaid.)

F. proserpinacoides, Willd.   Riverdale.   (Bicknell.)

Impatiens, L. (Jewel-weed—Touch-me-not.)

I. pallida, Nutt.   (Dr. Fisher.)
I. fulva, Nutt.   Damp grounds.

Oxalis, L. (Wood-Sorrel—Sheep-Sorrel.)

O. acetosella, L.   In damp woods, not common.
O. violacea, L.   Along fences and rocky places.
O. stricta, L.   Along fences and cultivated grounds, common.

Order 25.—RUTACEŒ. (Rue Family,)
Zanthoxylum, Colden. (Prickly Ash.)

Z. Americanum, Mill,   Rare.

Order 26—ANACARDIACEÆ, (Cashew Family,)

Rhus, L. (Sumach.)

R. typhina, L.   Frequent on rocky hill-sides.
R. glabra, L.   Common along fences.
R. copallina, L.   Hill-sides.   (Leggett.)
R. venenata, DC.   Swamps and damp grounds.
R. toxicodendron, L.   Common, especially along fences.

Order 27—VITACEŒ, (Vine Family,)
Vitis, Tourn.

V. Labrusca, L.   Damp thickets and woods.
V. æstivalis, Mx.   Thickets.   (Dr. Fisher.)
V. cordifolia, Mx.   Frost Grape, along streams.

Ampelopsis, Mx. (Virginian Creeper,)

    A. quinquefolia, Mx. Common in rich grounds.

Order 28—RHAMNACEÆ, (Buckthorn Family.)

Ceanothus, L. (New Jersey Tea—Red-root.)

    C. Americanus, L. Copses and borders of open woods.

Order 29—CELASTRACEÆ, (Staff-tree Family,)

Celastrus, L. (Staff-tree—Bitter-sweet.)

    C. scandens, L. Frequent throughout.

Euonymus, Tourn. (Burning Bush.)

    E. atropurpureus, Jacq. About White Plains.

    E. Americanus, L. Riverdale. Not common. (Bicknell.)

Order 30—SAPINDACEÆ, (Soapberry Family,)

Staphylea, L. (Bladde-nut.)

    S. trifolia, L. Throughout, sparingly.

Acer, Tourn. (Maple.)

    A. Pennsylvanicum, L., (Striped Maple.)

    A. spicatum, Lam. (Mountain Maple.)

    A. saccharinum, Wang, (Sugar Maple.) This is a favorite shade tree, on account of the well proportioned head it forms, and the beauty of its foliage. It is claimed that it is also a *fever tree*, *i. e.* When growing it absorbs so much water as to render malarial districts heathful. In the northern States sugar is manufactured from its sap.

    A. dasycarpum, Ehrhart, (White Maple—Silver Maple.) This tree was no doubt introduced from the west. On account of its rapid growth, it is a favorite street tree. It however forks in such a way as to be very liable to damage from high winds.

    A. rubrum, L. (Red Maple.) Common in swamps, bears transplanting to upland and is used sparingly for a shade tree.

Negundo, Mœnch. (Ash-leaved Maple—Box--Elder.)

    N. aceroides, Mœnch, A good shade tree, though requiring care to prevent a straggling habit of growth. Growing without cultivation about dwellings.

Order 31.—POLYGALACEÆ. (Milk Wort Famly,)

Polyagla. Tourn.

    P. lutea, L. New Castle, not common. (Hexamer.)

    P. sanguinea, L. Not rare.

    P. verticillata, L. Throughout these limits. NOTE.—There is reason to believe that several other species of this genus grow in the County.

Order 32.—LEGUMINOSÆ (Pulse Famly,)

Crotalaria, L.

   C. sagittalis, L.   Road-sides, frequent.

Trifolium, L.   (Clover.)

   T. arvense, L.   Sterile fields, common.   (Nat. Eu.)
   T. pratense, L.   Red clover.   (Adv. Eu.)
   T. repens, L.   White clover, common.
   T. agrarium, L.   Sandy fields and waste places.   (Nat. Eu.)
   T. procumbens, L.   Road-sides.   (Nat. Eu.)

Melilotus, Tourn   (Sweet Clover.)

   M. officinalis, Wild.   Waste grounds.   (Adv. Eu.)
   M. alba, Lam.   (Adv. Eu.)

Medicago, L.   (Medic.)

   M. lupulina. L.,   Waste lands.   (Adv. Eu.)

Robinia, L.   Locust Tree.

   R. Pseudacacia, L.   This tree is found in all parts of the county, but in late years it has not grown well; it is attacked by a borer, and dies at the top from some other cause.   (Nat. from the southwest.)

   R. viscosa, Vent.   Planted grounds, and found growing without cultivation near old dwellings.   (From the southwest.)

Astragalus, L.   (Milk Vetch.)

   A. Canadensis, L.   Rare.

Desmodium, DC.   (Tick Weed.)

   D. nudiflorum, DC.   Throughout.
   D. acuminatum, DC.   Throughout.
   D. pauciflorum, DC.   Not common.
   D. rotundifolium, DC.   Frequent.
   D. canescens, DC.   Not common.
   D. cuspidatum, Torr. and Gray.   Frequent.
   D. viridiflorum, Beck.   Frequent in the middle of the county.
   D. Dillenii, Darlingt.   (Bicknell.)
   D. paniculatum DC.   Shady woods, common.
   D. Canaden e, DC.   Woods, common.
   D. rigidum, DC.   Hill sides throughout.
   D. ciliare, DC.   (Bicknell.)
   D. Marilandicum, Boot.   Thickets, common.

Lespedeiza, Mich.   (Bush Clover.)

   L. procumbens, Mx.   Not common.
   L. repens, Torr. and Gray.   Frequent, Southern exposures.
   L. violacea, Pers.   Not rare.
      Var. angustifolia.   Frequent.
   L. hirta, Ell.   Dry, rocky hill sides.
   L. capitata, Mx.   Rye, near the shore of the sound.

Vicia, Tourn. (Vetch.)

V. sativa, L. Near Peekskill. (Le Roy) (Adv. Eu.)
V. tetrasperma, L. About Peekskill. (Le Roy.) (Nat. Eu.)
V. hirsuta, Koch. About Peekskill. (Le Roy.) (Nat. Eu.)
V. Americana, Muhl. About Peekskill. (Le Roy.)

Lathyrus, L. (Everlasting Pea.)

L. maritimus, Bigelow. Coast, common.
L. palustris, L.
    var. myrtifolius, Gray. (Bicknell.)

Apios, Boerhaave. (Wild Bean—Ground-nut.)

A. tuberosa, Mœch. Shady woods and damp copses, common.

Phaseolus, L. (Kidney Bean.)

P. diversifolius, Pers. (Bicknell.)
P. perennis, Walt. Woods and copses.
P. helvolus, L. Not rare, sandy fence rows.

Amphicarpæa, Ell. (Hog Peanut.)

A. monoica, Nutt. Rich woods.

Baptisia, Vent. (False Indigo.)

B. tinctoria, R. Br. Common.

Cassia, L. (Senna.)

C. Marilandica, L. Not common ; the leaves are purgative.
C. Chamæcrista, L. Frequent, damp, shady soil.
C. nictitans, L. Common, dry, sandy soil.

Gleditschia, L. (Honey Locust.)

G. triacanthos, L. This tree has been introduced from Southwest, and propagates itself ; it is a good shade tree.

Order 33.—ROSACEÆ. (Rose Family.)

Prunus, Tourn. (Plum, Cherry, etc.)

P. Americana, Marshall. Sing Sing. (Dr. Fisher.)
P. maritima, Wang. Rye.
P. Pennsylvanica, L. Wood lands and fence rows.
P. serotina, Ehrhart. Fence rows, not rare.

Spiraea, L. (Meadow Sweet.)

S. salicifolia, L. Damp places, not rare.
S. tomentosa, L. Frequent in edges of meadows.

Agrimonia, Tourn. (Agrimony.)

A. Eupatoria, L. Frequent in edges of woods.
A. parviflora, Ait. Riverdale, rare. (Bicknell.)

Poterium, L. (Burnet.)

    P. Canadense, Gray. (Canadian Burnet.)

Geum, L. (Avens.)

    G. Virginianum, L. (Bicknell.)
    G. album, Gmelin. Borders of woods.
    G. strictum, Ait. Sing Sing, not common. **(Bicknell.)**
    G. rivale, L. North Salem. (S. B. Mead.)

Potentilla, L. (Five-finger.)

    P. Norvegica, L. Not common.
    P. Canadensis, L. Along fences, and in old fields, **common.**
    Var. simplex, Torr & Gray. Common.
    P. argentea, L. Sterile grounds.
    P. arguta Pursh. (Hall.)

Fragaria, Tourn. (Strawberry.)

    F. Virginiana, Ehrhart. Fields, and woods.
    F. vesca, L. Fields.
    F. Indica, L. Not common. (Adv. India.)

Rubus, Tourn. (Brier, etc.)

    R. odoratus, L. Woods and damp copses.
    R. strigosus, Mx. Edges of woods and fence rows.
    R. occidentalis, L. Along fences, and borders of **woods.**
    R. villosus, Ait. Pastures, fence rows and woods.
    R. Canadensis, L. Pasture fields and along fences.
    R. hispidus, L. Damp grounds.

Rosa, Tourn. (Rose.)

    R. Carolina, L. Damp grounds and meadows.
    R. lucida, Ehrhart. Edges of meadows.
    R. blanda, Ait. Rocky places and edges of meadows.
    R. rubiginosa, L. White Plains, (Nat Eu.)
    R. micrantha, Smith. White Plains. (Nat. Eu.)

Cratægus, L. (Thorn.)

    C. coccinea, L. Thickets.
    C. tomentosa, L. Frequent in thickets.

Pyrus, L. (Pear, apple, etc.)

    P. arbutifolia, Ait. Thickets. (Hexamer.)
    Var. erythrocarpa, Gray. Peekskill. (Le Roy.)
    P. Americana, DC. Rocky woods. (Hexamer.)

Amalanchier, Medic. (June-Berry.)

    A. Canadensis, T. & Gray.
    Var. Botryapium, T. & Gray. Woods.

Order 35.—SAXIFRAGACEÆ. (Saxifrage Family.)

Ribes, L. (Currant and Goosberry.)
    R. Cynosbati, L. About Peekskill. (Le Roy.)
    R. lacustre, Poir. About Peekskill. (Le Roy.)
    R. floridum, L. Woods.

Philadelphus, L.
    P. coronarius, L. Near Jerome Park. (Bicknell.) Escaped from planted grounds.·

Parnassia, Poir. (Grass of Parnassus.)
    P. Caroliniana, Mx. Damp grounds.

Saxafraga, Mx. (Saxifrage—Rock-breaker.)
    S. Virginiensis, Mx. Damp rocks.
    S. Pennsylvanica, L. Edges of wet grounds.

Heuchera, L. (Alum-root.)
    H. Americana, L. Shaded rocks.

Mitella, Tourn. (Bishop's cap.)
    M. diphylla, L. Shaded hill sides.

Tierella, L.
    T. cordifolia, L. Shaded rocks.

Chrysosplenium, Tourn. (Golden Saxifrage.)
    C. Americanum, Schwein. White Plains.

Order 36.—CRASULACEÆ. (Orpine Family.)

Penthorum, Gronor. (Ditch Stone-crop.)
    P. sedoides, L. Wet meadows.

Tillæa, L.
    T. simplex, Nutt. Along banks of streams.

Sedum, Tourn. (Stone-crop—live forever.)
    S. acre, L. (Leggett.)
    S. Telephium, L. Escaped from cultivation. (Nat. Eu.)

Order 37.—HAMAMELACEÆ. (Witch-Hazel Family.)

Hamamelis, L. (Witch-Hazel.)
    H. Virginica, L. Damp woods, frequent.

Liquidamber, L. (Sweet-Gum –Bilsted.)
    L. Styraciflua, L. Near Pelham.

Order 38.—HALORAGEÆ. (Water-Milfoil Family.)

Myriophyllum, Vail. (Water-Milfoil.)
M. tenellum, Big. Wet places.

Proserpinaca, L. (Mermaid-weed.)
P. palustris, L. Swamps.

Order 39.—ONAGRACEÆ. (Evening-Primrose Family.)

Circæa, Tourn. (Enchanter's Nightshade.)
C. Lutetiana, L. Damp woods.

Epilobium, L.
E. angustifolium, L. Not rare.
E. palustre, L.
Var. lineare. Damp places, frequent.
E. coloratum, Muhl. Damp grounds.

Œnothera, L. (Evening Primrose.)
Œ. biennis, L. Common.
Œ. fruticosa, L. Kingsbridge. (Leggett.)
Œ. pumila, L. Rye Lake.

Ludwigia, L. (False Loose-strife.)
L. alternifolia, L. Common.
L. sphærocarpa, Ell. Peekskill. (Le Roy.)
L. palustris, Ell. Peekskill.

Order 41.—LYTHRACEÆ. (Loosestrife Family.)

Lythrum, L. (Loosestrife.)
L. Salicaria, L. Between Fordham and Williams Bridge. (Bicknell.)

Nesæa, Commerson, Juss. (Swamp Loose-strife.)
N. verticillata, H. B. K. Wet places.

Cuphea, Jacq.
C. viscosissima, Jacq., near Lake Mohegan.

Order 43.—CACTACEÆ. (Cactus Family.)

Opuntia, Tourn. (Prickly Pear.)
O. Rafinesquii, Eng. Rocks near New Rochelle.

Order 45.—CUCURBITACEÆ (Gourd Family.)

Sicyos, L. (One-seeded Star-Cucumber.)
S. angulatus, L. Damp grounds.

Order 46.—UMBELLIFERÆ. (Parsley Family.)

Hydrocotyle, Tourn. (Water Pennywort.)

H. Americana, L. Damp grounds.

H. umbellata, L. Ponds common.

Crantzia, Nutt.

C. lineata, Nutt. Coast, of the Sound.

Sanicula, Tourn. (Black Snakeroot.)

S. Canadensis, L. Shaded places.

S. Marilandica, L. Woods.

Daucus, Tourn. (Wild Carrot.)

D. Carota, L. Common. (Nat. Eu.)

Pastinaca, Tourn. (Parsnip.)

P. sativa, L. About dwellings, escaped from cultivation. It is generally supposed that it is poisonous, which is a mistake ; it is the Parsnip of the garden growing wild. (Adv. Eu.)

Archangelica, Hoffm.

A. hirsuta, Torr & Gray. (Leggett.)

A. atropurpurea, Hoffm. Along streams.

Conioselinum, Fischer. (Hemlock Parsley.)

C. Canadense, T. & G.

Æthusa, L. (Fools Parsley.)

Æ. Cynapium, L. (Adv. Eu.)

Thaspium, Nutt, (Meadow-Parsnip.)

T. aureum, Nutt. Sparingly scattered over the middle of the County.

Zizia, DC.

Z. integerrima, DC. Not common.

Discopleura, DC.

D. capillacea, DC. Salt marshes. (Leggett.)

Cuscuta, L. (Water-Hemlock.)

C. maculata, L. (Beaver Poison.) Northern part of the county.

C. bulbifera, L. Lake Mohegan. (Leggett.)

Sium, L. (Water-Parsnip.)

S. lineare, Mx. Wet places.

Cryptotænia, DC. (Honewort.)

C. Canadensis, DC. Thickets.

Osmorrhiza, Rof. (Sweet Cicely.)

O. longistylis, DC. (Damp woods.)

O. brevistylis, DC. (Woods and rich copses.)

Order 47.—ARALIACEÆ. (Ginseng Family.)

Aralia, Tourn. (Wild Sarsaprilla.

    A. spinosa, L. Near Van Cortland Lake. (Bicknell.)
    A. nudicaulis, L. (Bicknell.)
    A. racemosa, L. Shady woods.
    A. trifolia, Gray. Woods.

Order 48.—CORNACEÆ. (Dogwood Family.)

Cornus, Tourn (Dogwood.)

    C. florida, L. (Dogwood—Spoonwood.) A small tree. Common,
    C. circinata, L'Her. Not very frequent.
    C. sericea, L. Not rare.
    C. stolonifera, Mx. Edges of swampy places.
    C. paniculata, L. Her. Fence rows.
    C. alternifolia, L. Edges of copses, and along fences

Nyssa, L. (Sour Gum—Pepperidge.)

    N. multiflora, Wang. Damp woods and thickets.

Order 49.—CAPRIFOLIACEÆ. (Honeysuckle Family.)

Symphoricarpus, Dill. (Snowberry.)

    S. vulgaris, Mx. Rocky places. (Bicknell.)

Lonicera, L.

    L. sempervirens, Ait. Edges of woods.
    L. parviflora, Lam. Riverdale, rare. (Bicknell.)

Diervilla, Tourn. (Bush Honeysuckle.)

    D. trifida, Mœnch. Rocky places.

Triosteum, L. (Horse-Gentian.)

    T. perfoliatum, L. Edges of rich woods.

Sambucus, Tourn. (Elder.)

    S. Canadensis, L. Common.

Viburnum, F. (Arrow-wood.)

    V. Lentago, L. Along fences.
    V. prunifolium, L. (Bicknell,)
    V. nudum, L. Borders of swamps.
    V. dentatum, L. Damp grounds.
    V. acerifolium, L. Woods as undergrowth.
    V. pauciflorum, Pylaie.
    V. Opulus, L. Damp grounds. (S. B. Mead.'

Order 50.—RUBIACEÆ, (Madder Family.)

Galium, L. (Bedstraw—Cleavers,)
G. aparine, L. Moss-woods.
G. asprellum, Mx. Damp places.
G. trifidum, L. (Bicknell.)
G. triflorum, Mx. Woods.
G. pilosum, Ait. (Bicknell.)
G. circæzans, Mx. Open Woods.
G. lanceolatum, Tour. Woods.
G. boreale, L. Damp rocks.
G. verum, L. Peekskill. (Le. Roy.) (Adv. Eu.)

Cephalanthus, L. (Button-bush.)
C. occidentalis, L. Edges of meadows,

Mitchella, L. (Partridge-berry.)
M. repens, L. Shaded places.

Oldenlandia, Plumier, L.
O. glomerata, Mx. Damp places.

Houstonia, L.
H. purpurea, L.
H. cærulea, L. Among grass, not common.

Order 52.—DIPSACEÆ. (Teasel Family.)
Dipsacus, Tourn.
D. sylvestris, Mill, Road-sides. (Adv. Eu.)

Order 53.—COMPOSITÆ. (Composita Family.)

Vernonia, Schreb. (Iron-weed.)
V. Noveboracensis, Willd. Damp places, common.

Liatris, Schreb. (Blazing Star.)
L. scariosa, Willd. Rocky hills.)
L. spicata, Willd. Borders of salt meadows, Rye,

Eupatorium, Tourn. (Thoroughwort.)
E. purpureum, L. Damp grounds, common.
E. teucrifolium, Willd. Peekskill, (Le. Roy.)
E. sessilifolium, L. (Bicknell.)
E. perfoliatum, L. Damp grounds, common.
E. ageratoides, L. New Castle and Lake Mohegan.
E. aromaticum, L. Bedford.

Mikania, Willd.

    M. scandens, L.   Tarrytown.   (Paine.)

Tussilago, Tourn.   (Coltsfoot.)

    T. Farfara L.   Damp shaded banks.

Sericocarpus, Nees.   (White-topped Aster.)

    S. solidagineus, Nees.   Thickets, not rare

    S. conyzoides, Nees.   Dry copses.

Aster, L.

    A. corymbosus, Ait.   Woods, not rare.

    A. macrophyllus, L.   Damp shady woods.

    A. patens, Ait.   Along fences.

    A. lævis, L.   Copses, and hedge rows.

    A. undulatus, L.   Copses.

    A. cordifolius, L.   Road-sides and fence rows.

    A. sagittifolius, Willd.   Along fences.

    A. ericoides, L.   West Farms.

    A. multiflorus, Ait.   Road-sides.

    A. dumosus, L.   (Bicknell.)

    A. Tradescanti, L.   Damp grounds.

    A. miser. L., Ait.

    A. simplex, Willd.   Moist land.

    A. tenuifolius, L.   Damp grounds.

    A. longifolius, Lam.

    A. puniceus, L.   Common.

    A. Novæ-Angliæ, L.   Road-sides and meadows.

    A. acuminatus, Mx.   Peekskill.

    A. nemoralis, Ait.   Coast.

    A. flexuosus, Nutt.   (Bicknell.)

    A. linifolius, L.   New Castle.

Erigeron, L.   (Fleabane.)

    E. Canadense, L.   Very common.

    E. bellidifolium, Muhl.   Edges of woods, not rare.

    E. Philadelphicum, L.   Frequent.

    E. annuum, Pers.   Fields, not common.

    E. strigosum, Muhl.   New Castle, sparingly throughout.

Diplopappus, Cas.   (Double-bristled Aster.)

    D. linifolius, Hook.   Dry copses.

    D. umbellatus, T. & G.   Damp copses.

    D. amygdalinus, T. & G.   Cortlandt Lake.   (Bicknell.)

Solidago, L.   (Golden-rod.)

    S. squarrosa. Nutt.   Riverdale.   (Bicknell.)

    S. bicolor, L.   Hedge-rows,

    S. latifolia, L.   Damp shaded places.

S. cæsia, L.   Open woods.
S. speciosa, Nutt.
 Var. angustata.   (Bicknell.)
S. puberula, Nutt.   Sandy fields.
S. rigida, L.   Along fence rows.
S. sempervirens, L.   Rye, borders of salt marsh.
S. elliptica, Ait.   Borders of salt marsh, Rye.
S. neglecta, T. & G.   Wet places.
S. patula, Muhl.   Swampy places.
S. arguta, Ait.   Thickets.
 Var. juncea, G.   Thickets,
S. Muhlenbergii, T. & Gray.   Edges of woods.
S. linoides, Solander.   Swamps.
S. altissima, L.   Along fences.
S. ulmifolia, Muhl.   Damp thickets.
S. odora, Ait.   Peekskill.   (Le Roy.)
S. nemoralis, Ait.   Fields and road-sides, common.
S. Canadensis, L.   Fields and road-sides.
S. serotina, Ait.   Damp grounds.
S. gigantea, Ait.   Along fences.
S. lanceolata, L.   Damp grounds.
S. tenuifolia, Pursh.   Rye, near the coast.

Inula, L.   (Elecampane.)
 L. Helenium, L.   Road-sides, near dwellings.   (Nat. Eu.)

Pluchea, Cass.
 P. camphorata, DC.   Along the coast.

Baccharis, L.   (Groundsel-Tree.)
 B. halimifolia, L.   (Leggett.)

Iva, L.   (Marsh Elder.)
 I. frutescens, L.   Salt meadows.

Ambrosia, Tourn.   (Rag-weed.)
 A. trifida, L.   Fields.
 A. artemisiæfolia, L.   Said to be the plant whose pollen causes the disease
called Hay Fever.   In confirmation of this belief, the following statement was made
to me by Rev. Dr. Samuel Lockwood, who is a sufferer.   I give, as near as I can
recollect, his own words :   In a walk through the fields I came to a wheat stubble
covered with a dense growth of Ambrosia, artemisiæfolia, in a state just ready to
discharge its pollen.   I hesitated, but finally concluded to cross ; at once I found
great difficulty in breathing and when I reached the other side, I was completely
prostrated and was obliged to sit for some time, and the experiment was followed by
one of the most severe attacks of Hay Fever I have ever experienced.

Xanthium, Tourn.   (Cocklebur.)
 X. strumarium, L.   Spyten Duyvil, (Leggett.)
  Var. echinatum, Gray.
 Y. spinosum, L.   Coast.   (Nat. from the south.)

Heliopsis, L. (Tick-seed.)

H lævis, Pers.

Var. scabra. White Plains. A single plant in the nothern part of Scarsdale, near White Plains, found by Miss A. McCabe.

Rudbeckia, L. (Cone-flower.)

R. laciniata, L. Fields, sparingly throughout.

R. hirta, L. Fields, frequent.

Helianthus, L. (Sun-flower.)

H. giganteus, L. Swampy places.

H. strumosus, L. Damp thickets.

H. divaricatus, L. Edges of woods.

H. decapetalus, L. Along streams.

H. tuberosus, L. Escaped from cultivation. The tubers of this species are used for pickles.

Coreopsis, L. (Tick-seed.)

C. tricosperma, Mx. Wet places.

Bidens, L. (Bur-Marigold—Pitchfork.)

B. frondosa. L. Damp places.

B. connata, Muhl. Road-sides.

B. chrysanthemoides, Mx. New Castle.

B. bipinnata, L. Road-sides and fence rows.

Helenium, L. (Sneeze-weed.)

H. autumnale, L. Meadows.

Galinsoga, Ruiz & Par.

G. parviflora, Car. (Adv., from S. America.)

Maruta, Cass. (May-weed.)

M. Cotula, DC. Road-sides and wastes about dwellings, common. (Nat. Eu.)

Anthemis, L. (Chamomile.)

A. arvensis. White Plains. (Adv. Eu.)

Achillea, L. (Yarrow.)

A. Millefolium, L. Road-sides, Common.

Var. rosea.

Leucanthemum, Tourn. (Ox-eye Daisy.)

L. vulgare, Lam. (Nat. Eu.) White Plains is said to have been named from the appearance of the fields when this plant was in flower. (Doubtful.)

Tanacetum, L. (Tansy.)

T. vulgare, L. Near dwellings, (Adv. Eu.)

Artemisia, L. (Wormwood.)

A. vulgaris, L. About dwellings. (Adv. Eu.)

A. biennis, Willd. Along H. R. R. Road, not common.

Gnaphalium, L. (Cudweed.)

G. decurrens, Tow. Hills, frequent.

G. polycephalum, Mx. Edges of wood land.

G. uliginosum, L. Road-sides, throughout.

G. purpureum, L. Common in sterile soil.

Antennaria, Gærtin. (Everlasting.)

A. margaritacea, R. Brown. Woods.

A. platiginifolia Hook. Sterile hill sides.

Filago, Tourn. (Cotton-Rose.)

F. Germanica, L. Old fields. (Nat. Eu.)

Erechthites, Raf. (Fireweed.)

E. hieracifolia, Raf. Clearings, which have been burned over.

Seneceo, L. (Groundsel.)

S. vulgaris, L., (Waste places.) (Adv. Eu.)

S. aureus, L. Common.

Centaurea, L.

C. Cyanus, L. (Pooley.)

Cirsium, Tourn.

C. lanceolatum. Scop. Road-sides, (Nat. Eu)

C. discolor. Spreng. Damp thickets.

C. muticum, Mx. Wet grounds.

C. pumilum, Spreng. Old fields.

C. horridulum, Mx. Rye, and along the coast.

C. arvense, Scop, (Common in cultivated grounds.) This plant is found throughout, but not sufficiently abundant to be very troublesome. It fruits, if at all, very sparingly in this region.

Lappa, L. (Burdock.)

L. officinalis, Allioni. About dwellings.

Lampsana, L.

L. communis, L. Riverdale. (Bicknell—Adv. Eu.)

Cichorium, Tourn. (Chickory.)

C. Intybus, L. Road-sides and about dwellings. The root of this plant is used to adulterate coffee.

Krigia, Schreber. (Dwarf Dandelion.)

K. Virginica, Willd. Sparingly scattered over the county.

Cynthia, Don.

C. Virginica, Don. Not common.

Hieracium, Tourn. (Hawkweed.)

    H. Canadense, Mx. New Castle.

    H. scabrum, Mx. Dry copses and open woods.

    H. Gronovii, L.

    H. venosum, L.. Dry open woods, White Plains.

    H. paniculatum, L. Woods, not very common.

Nabalus, Cass. (Rattlesnake-root.)

    N. altissimus, Hook. Damp woods.

    N. Fraseri, DC. New Castle.

    N. racemosus, Hook. Hudson R. R. R. track, The seeds of this plant, have no doubt been brought from the west in grain, or other freight.

Taraxacum, Haller. Dandelion.

    T. Dens-leonis, Desf. Road-sides and lawns, common.

Lactuca, Tourn. (Lettuce.)

    L. Canadensis, L.

    Var. integrifolia, T. & G. (Leggett.)

    Var. sanguinea, T. & G. Dry grounds.

Mulgedium. Cass. (Blue Lettuce.)

    M. leucophæum, DC. Damp grounds.

Sonchus, L. (Sow-Thistle.)

    S. oleraceus, L.. near dwellings. (Nat. Eu.)

    S. asper, Vill. About dwellings. (Nat. Eu.)

    S. arvensis, L. Road-sides. (Nat. Eu.)

Order 54.—LOBELIACEÆ. (Lobelia Family.)

Lobelia, L, (Lobelia.)

    L. cardinalis, L.. Damp grounds. Very showy plant, bears cultivation.

    L. syphilitica, L. Road-sides and damp places ; bears cultivation well.

    L. inflata, L. Fields and pastures. Used by the Thomsonian practitioners in compounding their medicines.

    L. spicata, Lam. White Plains. (Miss McCabe.)

    L. Kalmii, L. White Plains. (Miss McCabe.)

Order 55.—CAMPANULACEÆ. (Campanula Family.)

Campanula, Tourn. (Bellflower.)

    C. rotundifolia, L. Leggett.

    C. aparanoides, Mx. Damp meadows, among grass.

    C. rapunculoides, L. (Bicknell.) (Adv. Eu.)

Specularia, Heister. (Venus's Looking-glass.)

    S. perfoliata, A. DC. Frequent.

Order 56.—ERICACEÆ. (Heath Family.)

Gaylussacia, II. B. K. (Huckleberry.)

G. dumosa, T. & G. Damp sandy soil.
G. frondosa, T. & G. Damp thickets.
G. resinosa, T. & G. Woodlands and swamps.

Vaccinium, L.

V. macrocarpon, Ait. The cranberry of commerce. Cranberry Pond near Kensico.
V. stamineum, L. Dry copses and open woods.
V. Pennsylvanicum, Lam. Dry copses and edges of woods.
V. vacillans, Solander. sandy woods.
V. corymbosum, L.
 Var. atrococcum, Gray. (Bicknell.)

Epigæa, L. (Trailing Arbutus.)

E. repens, L. Cranberry Pond, and valley of the Bronx, near the sound.—Somers near Coter's Lake, (James Wood.)—Sing Sing, (Dr. Fisher.)

Gaultheria, Kalm. (Wintergreen.)

G. procumbens, L. Damp woods.

Leucothoe, Don.

L. racemosa. Gray. (Bicknell.)

Cassandra, Don. (Leather leaf.)

C. calyculata, Don. New Castle. Bogs.

Andromeda, L.

A. Mariana, L. Along the coast of the Sound.
A. ligustrina, Muhl. White Plains,

Clethra, L. (Sweet pepperbush.)

C. alnifolia, Damp grounds and borders of meadows, bears transplanting and is used for an ornamental shrub.

Kalmia, L. (American Laurel.)

K. latifolia, L. Woods common, four to ten feet high, a beautiful evergreen shrub, bears transplanting and is valued for its heavy green leaves and showy flowers.
K. angustifolia. L. New Castle.

Azalia, L. (False Honeysuckle.)

A. viscosa, L. Not common. (Leggett.)
A. nudiflora, L. Woods throughout.

NOTE.—*Both the above species bear transplanting, and are beautiful objects in plant ed grounds. Shrubs three to six feet in height.*

Rhodora, Duhamel.
  R. Canadensis. Damp woods.

Pyrola, Tourn. (Shin-leaf.)
  P. rotundifolia. Boggy places.
    Var. asarifolia, Gray. not rare.
  P. elliptica, Nutt. White Plains.
  P. secunda, L. Woods.

Chimaphila, Pursh.
  C. umbellata. Nutt. Woods common.
  C. maculata, Pursh. Woods with the above.

Pterospora, Nutt. (Pine-drops.)
  P. Andromedea, Nutt.

Monotropa, L. (Indian Pipe.)
  M. uniflora, L. Dark woods.
  M. Hypopitys, L. Woods.

Order 55.—AQUIFOLIACEÆ. (Holly Family.)
Ilex, L. (Holly.)
  I. verticillate, Gray. (Black Alder.) Damp copses.
  I. lævigata, Gray. (Smooth winterberry.)

Order 59.—IBENACEÆ. (Ebony Family.)
Diospyros, L. (Persimmon, Medler.)
  D. Virginiana, L. Pelham. (Richard S. Collins.) A second-class tree.
Fruit edible, rare in these limits; no doubt introduced by seeds from further south.

Order 62.—PLANTAGINACEÆ. (Plantain Family.)
Plantago, L. (Plantain.)
  P. major, L. (Nat. Eu.)
  P. Rugelii. Decaisne. (Leggett.) Common.
  P. lanceolata, L. Pastures and lawns. (Nat. Eu.)
  P. Virginica, L. Sandy ground.
  P. pusilla, Nutt. Dry grounds.

Order 63.—PLUMBAGINACEÆ. (Leadwort Family
Statice, Tourn. (Marsh-Rosemary.)
  S. Limonium, L. Salt meadows along the Sound.

Order 64.—PRIMULACEÆ. (Primrose Family.)

Trientalis, L. (Chickweed-Wintergreen.)
  P. Americana, Pursh. Shady woods.

Steironema, Rof.
  S. ciliata, L. Low grounds.
  S. lanceolata, Walt. Wet banks of streams.

Lysimachia, Tourn. (Loosestrife.)
  L. thyrsiflora, L. Swampy places.
  L. stricta, Ait. Damp places.
  L. quadrifolia, L. Damp ground.
  L. nummularia, L. Escaped from cultivation. (Leggett.)

Anagallis, Tourn. (Pimpernel.)
  A. arvensis, L. Old fields. (Nat. Eu.)

Samolus, L.
  S. Valerandi, L.
    Var. Americanus, Gray, (Leggett.)

Hottonia, L. (Water Violet.)
  H. inflata, Ell. Stagnant water.

Order 65.—LENTIBULACEÆ. (Bladderwort Family.)

Utricularia, L. (Bladderwort.)
  U. vulgaris, L.
  U. gibba, L. Shoal water.

Order 66.—BIGNONIACEÆ. (Bignonia Family.)

Catalpa, Scop., Walt. (Catalpa. Indian Bean. Smoking Bean Tree.)
  C. bignonioides, Walt. This tree is rather a favorite as a shade tree on account of its fine foliage and showey flowers. Introduced from the West, and sows itself and grows without cultivation along the road sides. In late years nurseymen graft it, and thereby produce a more compact head.

Order 67.—OROBANCHACEÆ. (Broom-rape Family.)

Epiphegus, Nutt. (Beech-drops.)
  E. Virginiana, Bart. Shady woods.

Conopholis. Wallroth. (Squaw-root.)
  C. Americana, Wallroth. Woods.

Aphyllon, Mitchell. (Naked Broom-rape.,
  A. uniflorum, T. & G. Woods.

Order 68.—SCROPHULARIACEÆ. (Figwort Family.)

Verbascum, L. (Mullein.)
  V. Thapsus, L. Fields common. (Nat. Eu.)
  V. Blattaria, L. Fields and road sides. (Nat. Eu.)

Linaria, Tourn. (Toad Flex. Snap Dragon.)
  L. Canadensis, Spreng. About Peekskill. (Le Roy.)
  L. vulgaris, Mill. Fields and pastures, a troublesome weed. (Nat. Eu.)

Scrophularia, Tourn. (Figwort.)
  S. nodosa, L. Damp thickets.

Chelone, Tourn. (Turtle-head. Snake-head.)
  C. glabra, L. Borders of wet meadows.

Pentstemon, Mitchell. (Beard-tongue.)
  P. pubescens, Solander. Edges of thickets.

Mimulus, L. (Monky-flower.)
  M. ringens, L. Edges of meadows and wet thickets.

Gratiola, L. (Hedge-Hyssop.)
  G. Virginiana, L. Banks or shores of ponds.

Ilysanthes, Raf.
  I. gratioloides, Benth. (Bicknell.)

Limosella, L. (Mudwort.) ·
  L. aquatica, L.
    Var. tenuifolia, Hoffm. Along the coast.

Veronica, L. (Speedwell.)
  V. Virginica, L. Woods.
  V. Americana, Schweinitz. Along the edges of ditches.
  V. scutellata, L. Swamps.
  V. officinalis, L. Dry grounds.
  V. serpyllifolia, L. Road sides and lawns.
  V. peregrina, L. Waste places.
  V. arvensis L. Cultivated grounds.

Gerardia, L.
  G. purpurea, L. Damp places.
  G. maritima, Raf. Rye and along the coast.
  G. tenuifolia, Vohl. Woods.
  G. flava, L. Woods and copses.
  G. quercifolia, Pursh. Woods.

Castilleia, Mutis. Painted-cup.)
  C. coccinea, Spreng.

Pedicularis, Tourn. (Lousewort.)

    P. Canadensis, L. Copses.
    P. lanceolata, Mx. Edges of wet meadows.

Melampyrum, Tourn. Cow-wheat.

    M. Americana, Mx.

Order 70.—VERBENACEÆ. (Vervain Family.)

Verbena, L. (Vervain.)

    V. angustifolia, Mx. Sing Sing. (Dr. Fisher.)
    V. hastata, L. Waste places about dwellings.
    V. urticifolia, L. Road-sides and borders of fields.

Phryma, L.

    P. Leptostachya, L. Damp woods.

Order 71.—LABIATÆ. (Mint Family.)

Teucrium, L. (Germander.)

    T. Canadense, L. Damp places.

Trichostema, (Blue Curls.)

    T. dichotomum, L. Common in stubble.

Mentha, L. (Mint.)

    M. viridis, L. Damp places. (Nat. Eu.)
    M. piperita, L. Along brooks. (Nat. Eu.)
    M. aquatica, L.
      Var. crispa, Benthem. (Leggett.) (Nat. Eu.)
    M. arvensis, L. Peekskill. (Le Roy.)
    M. Canadensis, L. Peekskill. (Le Roy.)

Lycopus, L.

    L. Virginicus, L. Damp meadows.
    L. Europæus, L. (Bicknell.)

Cunila, L. (Dittany.)

    C. Mariana, L. Old fields.

Pycnanthemum, Mx. (Sweet Basil. Mountain mint.)

    P. incanum, Mx. Edges of woods.
    P. clinopodioides, T. & G. Copses.
    P. muticum, Pers. Along fences.
    P. lanceolatum, Pursh. Copses and along fences.
    P. linifolium, Pursh. Riverdale. (Bicknell.)

Origanum, Wild Marjoram.

    O. vulgare, L. Near dwellings, escaped from cultivation. (Nat. Eu.)

Thymus, I.. (Thyme.)

    T. Serpyllum, L.   Escaped from cultivation.   (Adv. Eu.)

Calamintha, Mœnch.

    C. Clinopodium, Benth.   Edges of thickets and along fence rows.

Melissa, L.. (Balm.)

    M. officinalis, L.   Near dwellings where it has escaped from cultivation. (Nat. Eu.)

Hedeoma, Pers.   (Mock Pennyroyal.)

    H. pulegioïdes. Pers.   Old fields and woods.

Collinsonia, L..   (Horse Balm.)

    C. Canadensis, L..   Damp shady woods.

Salvia, L.. (Sage.)

    S. lyrata, L..   New Castle.   (Hexamer.)

Monarda, L.   (Horse Mint.)

    M. didyma, L..   Peekskill.   Not common. **(Le Roy.)**
    M. fistulosa, L..   Old fields near Little Rye Lake.
    M. punctata, L..   Not common.   Sing Sing. **(Dr. Fisher.)**

Blephilia, Raf.

    B. ciliata, Raf.   Peekskill.   (Le Roy.)

Lophanthus, Benth.   (Giant Hyssop.)

    L. scrophulariæfolius, Benth.   Edges of woods.

Nepeta, L.. (Cat-Mint. Catnip.)

    N. Cataria, L..   Near dwellings, common.   (Adv. Eu.)
    N. Glechoma, Benth.   A weed in gardens.

Physostegia, Benth.   (False Dragon-Head.)

    P. Virginiana, Benth.   Not Common.

Brunella, Tourn.   (Self-Heal.)

    B. vulgaris, L..   Fields and open woods,

Scutellaria, L.. (Skullcap.)

    S. pilosa, Mx.   Jerome Park.   (Bicknell.)
    S. integrifolia, L..   Along fences.
    S. galericulata, L..   Shady places along brooks.
    S. lateriflora, L..   Damp places.

  Marrubium, L.. (Horehound.)

    M. vulgare, L..   About dwellings.   (Nat. Eu.)

Galeopsis, L.. (Hemp-Nettle.)

    G. Tetrahit, L..   Waste places.   (Nat. Eu.)

Stachys, L.
  S. palustris, L.
    Var. aspera, Gray. Not common. (Leggett.)

Leonurus, L. (Motherwort.)
  L. Cardiaca, L. Near dwellings. (Nat. Eu.)

Lamium, L. (Dead-Nettle.)
  L. amplexicaule, L. Weed in gardens. (Adv. Eu.)

Order 72.—BORRAGINACEÆ. (Borage Family.)

Echium, Tourn. Viper's Bugloss.
  E. vulgare, L. Road-sides, not common. (Nat. Eu.)

Symphytum, Tourn. (Comfrey.)
  S. officinale, L. Road-sides, escape from cultivation. (Adv. Eu.

Onosmodium, Mx. (False Gromwell.)
  O. Virginianum, DC. Hill sides.

Lithospermum, Tourn. (Gromwell. Puccoon.)
  L. officinale, L. Peekskill.

Myosotis, L. (Forget-me-not.)
  M. palustris, Withering.
    Var. laxa, Gray. Along brooks and other wet places.
  M. verna, Nutt. (Bicknell.)

Cynoglossum, Tourn. (Hound's-Tongue.)
  C. officinale, L. Damp places and fields. (Nat. Eu.)
  C. Morisoni, DC. (Pooley.)

Ord r 73.—HYDROPHYLLACEÆ. (Waterleaf.)

Hydrophyllum, L. (Waterleaf.)
  H. Virginicum, L. Shady woods.

Order 74.—POLEMONIACEÆ. (Polemonium Family.)
Phlox, L.
  P. subulata, L. Sing Sing. (Dr. Fisher.)

Order 75.—CONVOLVULACEÆ. (Convolvulus Family.)

Quamoclit, Tourn. (Cypress-Vine.)
  Q. coccinea, Mœnch. Banks of stream. (Nat. Trop. America.)

Ipomœa, L. (Morning-glory.)
> **I.** purpurea, Lam. Near dwellings. (Adv. from South.)
> **I.** pandurata, Meyer. Spuyten-Duyvil. (Bicknell.)
> **I.** Nil, Roth. (Bicknell.) (Adv. Trop. America.)

Convolvulus, L. (Bindweed.)
> **C.** arvensis. L. Sing Sing. (Dr. Fisher.) (Nat. Eu.)

Calystegia, R. Br. (Bindweed.)
> **C.** sepium, R. Br.

Cuscuta, Tourn. (Dodder.)
> **C.** inflexa, Eng. Damp places, Hudson R. R. track, and Lake Mohegan.
> **C.** Gronovii, Willd, Common in damp grounds.
> **C.** compacta, Juss. Damp places, Hudson R. R. track.

Order 76.—SOLANACEÆ. (Nightshade Family.)

Solanum, Tourn. (Nightshade.)
> **S.** Dulcamara, L. White Plains. (Nat. Eu.)
> **S.** nigrum, L. West Farms. (Nat. Eu.)

Physalis, L. (Ground cherry. Ground apple.)
> **P.** pubescens, L. Peekskill. (Le Roy.)
> **P.** viscosa, L. Road-sides, common.

Nicandra, Adans. (Apple of Peru.)
> **N.** physaloides, Gærtn. Waste grounds. (Adv. Peru.)

Lycium, L.
> **L.** vulgare, Dunal. Matrimony-vine. Near dwellings, rare. (Adv. Eu.)

Hyoscyamus, Tourn. (Henbane.)
> **H.** niger, L. About dwellings. (Adv. Eu.)

Datura, L. (Simon Pumpkin. Thorn apple.)
> **D.** Stramonium. About dwellings. (Adv. Asia.,
> **D.** Tatula, L. (Pooley.)

Order 77.—GENTIANACEÆ. (Gentian Fam..y.)

Sabbatia, Adans. (American Centuary.)
> **S.** angularis, Pursh. (Bicknell.)
> **S.** stellaris, Pursh. Salt marshes. (Leggett.)
> **S.** chloroides, Pursh. Along the coast.

Gentiana, L. (Gentian.)

G. crinita, Frœl. Low grounds.

G. Andrewsii, Griseb. Damp grounds.

Bartonia, Muhl.

B. tenella, Muhl. Woods. (Leggett.)

Menyanthes, Tourn. (Buckbean.)

M. trifoliata, L. Cranberry Pond. (Kensico.)

Order 79.—APOCYNACEÆ. (Dogbane Family.)

Apocynum, Tourn. (Indian Hemp.)

A. androsæmifolium, L. Edges of thickets.

A. cannabinum, L. Not rare.

Order 80.—ASCLEPIADACEÆ. (Milkweed Family.)

Asclepias, L. (Milkweed. Silkweed.)

A. Cornuti. Decaisne. Common.

A. phytolaccoides, Pursh. New Castle.

A. purpurascens, L. Copses and edges of woods.

A. variegata, L. Woods.

A. quadrifolia, Jacq. Shady woods, frequent.

A. incarnata, L. Damp places.

A. tuberosa, L. Dry fields and road-sides.

A. verticillata, L. Not rare.

Acerates, Ell. (Le Roy.) (Green Milkweed.)

A. viridiflora, Ell. Not rare.

Order 81.—OLEACEÆ. (Olive Family.)

Ligustrum, Tourn. (Privet, or Prim.)

L. vulgare, L. A graceful shrub, growing frequently without cultivation, used for hedges. (Nat. Eu.)

Fraxinus, L.

F. Americana, L. (White Ash.) Throughout; a large tree used for ornamental purposes. The wood is strong, is used for oars, and also for flooring.

F. pubescens, Lam. (Red Ash.) A middle size tree, not common in these limits. Sing Sing. (Dr. Fisher.)

F. sambucifolia, Lam. (Black Ash.) Much used in the manufacture of strong baskets.

Order 82.—ARISTOLOCHIACEÆ. (Birthwort Family.)

Asarum, Tourn. (Wild Ginger.)

A. Canadense, L. White Plains.

Aristolochia, Tourn. Birthwort.

    A. serpentaria, L. (White snake root.) Not common.

    A. Sipho, L. Her. (Dutchman's Pipe.) Planted for ornament and escaped, near dwellings. (From south-west.)

Order 84.—PHYTOLACCACEÆ. (Pokeweed Family.)

Phytolacca, Tourn. (Pokeweed.)

    P. decandra, L. Rich grounds and clearings.

Order 85.—CHENOPODIACEÆ. (Goosefoot Family.)

Chenopodium, L. (Pigweed.)

    C. album, L. A weed in gardens. (Nat. Eu.)

    C. urbicum, L. Cultivated grounds. (Nat. Eu.)

    C. murale, L. Peekskill. (Adv. Eu.)

    C. hybridum, L. About dwellings. (Nat. Eu.)

    C. Botrys, L. West Farms. (Adv. Eu.)

    C. ambrosioides, L. Waste grounds about dwellings. (Nat. Trop. America.)

    NOTE.— *The last two species are used in medicine as a vermifuge.*

Blitum, Tourn.

    B. maritimum, Nutt. Rye.

    B. Bonus-Henricus, Reichenbach. About dwellings. (Adv. Eu.)

Atriplex, Tourn. (Orache.)

    A. patula, L.

    Var. hastata, Gray. Rye, and coast.

    A. arenaria, Nutt. Rye, and along the coast.

Salicornia, Tourn. (Glasswort. Samphire.)

    S. herbacea, L. Rye, and along the coast.

Suæda, Forskal. (Sea Blite.)

    S. maritima, Dumortier, Coast.

Salsola, L. (Saltwort.)

    S. Kali, L. Coast of Long Island Sound.

Order 86.—AMARANTACEÆ. (Amaranth Family.)

Amarantus, Tourn. (Amaranth.)

    A. hypochondriacus, L. Sing Sing. (Dr. Fisher. Trop. America.)

    A. paniculatus, L. Sing Sing. (Dr. Fisher. Trop. America.)

    A. retroflexus, L. Weed in gardens.

    A. albus, L. Wast places near dwellings. (Adv. Trop. America.)

    A. spinosus, L. About dwellings. (Adv. Trop. America.)

    A. pumilus, Raf. Near Rye. (Mead.)

    A. viridis, L. Along Hudson River R. R. track.

Acnida, L. (Water Hemp.)

    A. cannabina, L.

Order 87.—POLYGONACEÆ. (Buckwheat Family.)

Polygonum, L. (Knotweed.)

    P. orientale, L. About dwellings. (Adv. India.)

    P. Careyi, Olney. (Bicknell.)

    P. Pennsylvanicum, L. Damp waste places.

    P. incarnatum, Ell. (Bicknell.)

    P. Persicaria, L. Waste places ; Commom.

    P. Hydropiper, L. Moist grounds.

    P. acre, H. B. K. Wet grounds, common.

    P. hydropiperoides, Mx. Wet places.

    P. amphibium, L. Water and very wet places.

    P. Virginicum, L. Thickets and road-sides.

    P. aviculare, L.

        Var. erectum, Roth. About dwellings with the last, **but larger.**

    P. maritimum, L. Coast of Long Island Sound.

    P. tenue, Mx. Dry soil.

    P. arifolium, L. Wet grounds.

    P. sagittatum, L. Damp grounds.

    P. Convolvulus, L. Cultivated fields. (Nat. Eu.)

    P. cilinode, Mx. Rocks and thickets.

    P. dumetorum, L. Damp places.

        Var. scandens, Gray. Thickets.

Fagopyrum, Tourn. (Buckwheat.)

    F. esculentum, Mœnch. Escaped from cultivation. (Adv. Eu.**)**

Rumex, L. Dock. (Sorrel.)

    R. Britannica, L. Damp grounds.

    R. crispus, L. (Curled Dock.) Common in grass fields, used **for greens, a** troublesome weed. (Nat. Eu.)

    R. obtusifolius, L.

    R. Acetosella, L. (Horse Sorrel. Field Sorrel.) Troublesome weed **in old** fields. (Nat. Eu.)

Order 88.—LAURACEÆ. (Laurel Family.)

Sassafras, Nees.

    S. officinale, Nees. Second-class tree in woods.

Lindera, Thunberg. (Wild Allspice. Spice-wood.)

    L. Benzoin, Meisner. Shady woods.

Order 89.—THYMELEACEÆ.   (Mezereum Family.)

Dirca, L.   (Moose-wood. Leatherwood.)

> **D.** palustris, L.   Woods, used for ornamental purposes ; small tree

Order 91.—SANTALACEÆ.   (Sandalwood Family.)

Comandra, Nutt.   (Bastard Toad-flax.)

> **C.** umbellata, Nutt.   Not rare.

Order 93.—SAURURACEÆ.   (Lizard's-Tail Family.)

Saururus, L.   (Lizard's-tail.)

> **S.** cernuus, L.   Edges of sluggish streams and pools.

Ceratophyllum, L.

> **C.** demersum, L.

Order 97.—EUPHORBIACEÆ.   (Spurge Family.)

Euphorbia, L.   (Spurge.)

> **E.** polygonifolia, L.   Shores of the Sound.
> **E.** maculata, L.   Road-sides, common.
> **E.** hypericifolia, L.   A weed. Fields and gardens.
> **E.** Cyparissias, L.   About dwellings. (Adv. Eu.)

Acalypha, L.   (Three-seeded Mercury.)

> **A.** Virginica, L.   A weed in gardens and waste places about dwellings.

Order 99. URTICACEÆ,   (Nettle Family.)

Ulmus, L.   (Elm.)

> **U.** fulva, Mx. (Slippery Elm.) This tree is well-known, the inner bark is charged with mucilage, and is used largely for poultices, and as a remedy in throat diseases ; from 30 to 40 feet high.
>
> **U.** Americana, L. (American Elm.) A large tree reaching the height of 80 to 90 feet. On account of the graceful forking of its branches, it is highly valued for ornamental purposes ; when planted in rows along avenues, the forked branches interlace, forming pointed arches.
>
> **U.** racemosa, Thomas. (White Elm. Corky Elm.) A large tree, not so desirable as the last for a shade or ornamental tree, on account of the roughness of its branchlets which are flanked by corky wings.

Celtis, Tourn.   (Nettle-tree. Hackberry.)

> **C.** occidentalis, L. (Hackberry.)   River banks.

Morus, Tourn. (Mulberry.)

M. rubra, L. (Red Mulberry.) Fields and hedge-rows, a small tree from 20 to 30 feet high, bearing edible fruit ; the timber is hard and durable, the roots are much used for the knees of rowboats and skiffs.

M. alba. L. (White Mulberry.) This tree grows taller than the last, reaching the height of 50 feet; the timber is soft and brash, the fruit is white and edible; but insipid and less desirable than the last.

Urtica, Tourn. (Nettle.)

U. gracilis Ait. Waste, damp grounds.

U. dioica, L. Waste grounds about dwellings. (Nat. Eu.)

U. urens, L. Waste grounds near dwellings. (Nat. Eu.)

Laportea, Gaudichaud. Damp woods.

L. Canadensis, Gaudichaud. Damp woods.

Pilea, Lindl. (Clearweed.)

P. pumila, Gray. Shady places.

Bœhmeria, Jacq. (False Nettle.)

B. cylindrica, Willd. Damp shady places.

Cannabis, Tourn. (Hemp.)

C. sativa, L. Escaped from cultivation. (Adv. Eu.)

Humulus, L. (Hop.)

H. Lupulus, L. Damp rich copses.

Order 100.—PLATANACEÆ. (Plane-tree Family.)

Platanus, L. (Plane-tree. Buttonwood.)

P. occidentalis, L. (Sycamore.) This a large tree planted for ornamental purposes, especially in rows by the road-sides.

Juglans, L.

J. cinerea, L. (Butter-nut.) Grows in the hills, and the rocks seem favorite places for it. The fruit is valuable, and the wood is used for cabinet purposes. It has a coarse grain but takes a good polish.

J. nigra, L. (Black walnut.) A large tree reaching the height of 80 to 100 feet ; planted about houses for its fruit. The wood is of a dark color and is much used for cabinet work.

Carya, Nutt. (Hickory.)

C. alba, Nutt. (Shell-bark Hickory. Shag-bark Hickory.) A large straight tree growing to the height of 40 to 70 feet. The fruit is highly valued on account of its excellence, and the thinness of the shell. The wood splits easily, and makes excellent fuel.

C. sulcata, Nutt. (Thick Shell-bark. Hickory.) Large tree reaching the height of 80 feet, the fruit is much larger than the fruit of C. Alba. It is not common in this region, and has undoubtedly sprung from seed brought from the West. The wood and bark is much like those of C. Alba.

C. tomentosa, Nutt. (White-heart Hickory. Bull-nut.) A large tree, nuts with very thick shells. Wood with straight grain, splits well, and makes good fuel.

C. porcina, Nutt. (Pig-nut. Broom Hickory.) The fruit, not desirable. The wood is tough, and is used for axe and hammer handles, for hubs and spokes of wagon wheels. A large tree sometimes reaching the height of 100 feet, common.

C. amara, Nutt. (Bitter-nut or Swamp Hickory.) Large tree, reaching 40 to 60 feet in height ; wood not valuable for timber, though it makes good fuel. The fruit has a thin shell and is bitter, common.

Order 102.—CUPULIFERÆ. (Oak Family.)

Quercus, L. (Oak.)

Q. alba. L. (White Oak.) A large tree reaching the height of 75 feet, or more; and not unfrequently attains a diameter of 6 feet. A very valuable tree on account of the durability of its wood, common.

Q. obtusiloba, Mx. (Post-oak or Box White-oak.) Small tree, wood hard and durable, used for axe and pick handles, also for fence posts.

Q. macrocarpa, Mx. (Bur-oak. Mossy-cup. White-oak.) A middle sized tree, 40 to 50 high, forms symetrical head. Have not seen this tree in these limits. and put it down on the authority of Dr. Fisher, of Sing Sing.

Q. bicolor, Willd. (Swamp White-oak.) A fine tree, reaching the height of 70 feet. Its wood furnishes durable timber and excellent fuel.

Q. Prinos, L. (Chestnut-oak.) Straight growing tree of middle size in these limits; the wood makes excellent fuel.

Var. acuminata, Mx. (Yellow Chestnut Oak.) Leaves like those of the chestnut tree. A middle sized tree in these limits ; wood soft, used for fuel.

Q. coccinea, Wang. (Scarlet Oak.) Large tree, wood makes good fuel ; and the bark much used in tanning leather.

Var. tinctoria. (Yellow Bark Oak, Black Oak.) Large tree reaching sometimes to the height of 100 feet when growing in the forest. Wood used for fuel and the bark for tanning.

Q. rubra, L. (Red Oak.) Large tree, coarse grained, used for fuel.

Q. palustris, Du Roi. (Pin Oak, Spanish Oak, Water Oak.) Wet grounds ; reaching the height of 50 feet or more in these limits, and three to four feet in diameter ; timber hard and close, difficult to split.

Castanea, Tourn. (Chestnut.)

C. vesca. L. woods and fence rows. Common. A large tree, valued on account of its excellent fruit and durable timber ; much used for fencing, and in late years, largely for cabinet work ; the grain is coarse, but takes a good polish ; reaches the height of 90 feet.

Fagus, Tourn. (Beech.)

F. ferruginea, Ait. (American Beach.) Large tree, 40 to 70 feet in height; forms a graceful head, when growing separately, and on that account is a desirable tree for ornamentable purposes; wood makes excellent fuel.

Corylus, Tourn. (Hazel-nut.)

C. Americana, Walt. (Wild Hazel-nut.) Thickets and road-sides; fruit edible. A shrub 5 to 8 feet high.

C. rostrata, Ait. (Beaked Hazel-nut.) A shrub like the last; fruit edible.

Ostrya, Micheli.

O. Virginica, Willd. (Hop-Hornbeam. Iron-wood.) Second class tree, found in the borders of woods, and in copses; sometimes used for ornamental purposos.

Carpinus, L. (Hornbeam. Iron-wood.)

C. Americana, Mx. (American Hornbeam, Blue Beach, Water Beach.) A second class tree, reaching the height of 20 feet; used for ornamental purposes, and for hedging.

Order 103.—MYRICACEÆ. (Sweet Gale Family.)

Myrica, (Bayberry. Wax-Myrtle.)

M. cerifera, L. (Bayberry.) Shrub, sometime used for ornamental purposes, the pulucrized leaves used for snuff for Cattarrah; and the bark of the root enters largely into the Thompsonian remedies.

Comptonia, Solander. (Sweet-Fern.)

C. asplenifolia, Ait. Road-side, in sunny places.

Order 104.—BETULACEÆ. Birch Family.)

Betula, Tourn. (Birch.)

B. lenta, L. (Sweet Birch, Black Birch.) A large tree, making excellent fuel, and used also for cabinet work.

B. lutea, Mx. f. (Yellow Birch.) Large tree, sometimes reaching the height of 80 feet, and 3 to 4 feet in diameter; rare in these limits; damp cold woods; used in Nova Scotia, in ship building. The keel of the largest ship ever built in Nova Scotia, was Birch.

B. alba.

Var. populifolia, Spach. (White Birch.) A small slender tree.

B. nigra, L. (Red Birch.) Large tree growing along river banks, good for fuel.

Alnus, Tourn. (Alder.)

A. incana, Willd. (Speckled Alder.) A shrubby tree 15 to 18 feet high; wet places.

A. serrulata, Ait. (Smooth Alder.) Shrub 10 feet high; wet grounds and swampy places.

Order 105.—SALICACEÆ. (Willow Family.)

Salix, Tourn. (Willow.)

    S. candida, Willd. (Hoary Willow. Shrub 6 feet high.

    S. tristis, Ait. (Dwarf Gray Willow) Shrub 2 feet high ; wet places.

    S. discolor, Muhl, (Glaucous Willow.) Banks of brooks ; 15 feet high.

    S. sericea, Marshall. (Silkey Willow.) Banks of streams.

    S. viminalis, L. (Basket Willow.) Shrub growing in damp ground. (Adv. Eu.)

    S. cordata, Muhl. (Heart-leaved Willow,) Small tree. (Leggett.)

    S. livida, Wahl.

    Var. occidentalis, Grey. Shrub 10 feet high. (Leggett.)

    S. petiolaris, Smith. Small tree 15 to 20 feet high.

    S. lucida, Muhl. Along streams, 15 feet high.

    S. nigra, Marsh.

    Var. falcata, Gray. Tree 30 feet high.

    S. fragilis, L. Large tree, wet grounds, and river-banks. (Adv. Eu.)

    S. alba, L.

    Var. vitellina, Gray. (Yellow Willow.) A large tree, along streams and river-banks. (Adv. Eu.)

Populus, Tourn. (Poplar, Aspen.)

    P. tremuloides, Mx. (Aspan Leaf.) Large tree.

    P. grandidentata, Mx. Large tree, taller than the last.

    P. balsamifera, L. Balsam, Poplar. (Tacamahac.)

    Var. candicans, Gray. (Balm of Gilead.) Large tree.

    P. alba. L. (Abele Tree.) Shade tree, introduced from France, by nursery-men, sometimes called silver leaved poplar.

    Note.— *The whole genus suckers profusely, and is on that account objectionable for lawn purposes.*

Order 106.—CONIFERÆ. (Pine Family.)

Pinus, Tourn. (Pine.)

    P. rigida, Miller. (Pitch Pine.) Large tree, timber hard and filled with resin.

    P. inops, Ait. (Scrub Pine.) Small tree, wood hard but not large enough for valuable timber. Sing Sing, Dr. Fisher. (Must have been planted.)

    P. mitis, Mx. (Yelow Pine.) Large tree, good timber.

    P. Strobus, L. (White Pine.) Largest of the Genus, growing to the height of 150 feet, makes excellent boards and planks.

Abies, Tourn. (Spruce Fir.)

    A. nigra, Poir. (Black Spruce.) Large tree furnishing excellent timber. North Salem. (Dr. Mead.)

    A. Canadensis, Mx. (Hemlock.) Large tree, used for ornamental and Hedging purposes. Largely used for lumber.

Larix, Tourn. (Larch.)

L. Americana, Mx. (Larch, Black Larch, Hackmatack, Tamarack.) A large tree, used for ornamental purposes, and for lumber.

Thuja, Tourn. (Arbor Vitæ.)

T. occidentalis, L. (American Arbor Vitæ.) This tree is also called Cedar, in the north where it is used in the manufacture of barrels. It is a favorite ornamental tree ; largely used in hedges.

Juniperus, L. (Juniper.)

J. Virginiana, L. (Red Cedar.) A tree from 15 to 40 feet high in these limits ! wood is close-grained and takes a good polish ; makes durable fence posts.

CLASS II.—MONOCOTYLEDONOUS OR ENDOGENOUS PLANTS.

Order 107.—ARACEÆ, (Arum Family.)

Arisæma, Martius. (Indian Turnip.)

A. triphyllum, Torr. Damp places.

Peltandra, Raf.

P. Virginica, Raf. Shoal water, or very wet places. (Leggett.)

Calla, L. (Water Arum.)

C. palustris, L. Cold swampy land. Sing Sing, (Dr. Fisher.)

Symplocarpus, Salisb. (Skunk Cabbage.)

S. fœtidus, Salisb. Damp grounds ; common

Acorus, L. (Sweet Flag. Calamus.)

A. Calamus, L. Margins of small streams in swampy places.

Order 108.—LEMNACEÆ. (Duckweed Family.)

Lemna, L. (Duckweed. Duck's-meat.)

L. polyrrhiza, L. Ponds. (Leggett.)

Order 109.—TYPHACEÆ. (Cat-tail Family.)

Typha, Tourn. (Cat-tail Flag.)

T. latifolia, L. Wet places.

T. angustifolia, L. Wet places.

Sparganium, Tourn. (Bur-reed)

S. eurycarpum, Engelm. (Bicknell.)

S. simplex, Hudson. Not rare.

Order 110.—NAIDACEÆ.   (Pondweed Family.)

Naias, L.   (Naiad.)
    N. flexilis, Rostk.   (Bicknell.)

Zannichellia, Micheli.   (Horned Pondweed.)
    Z. palustris, L.   Lower part of the county, now New York.   (Bicknell.)

Zostera, L.   (Eel-grass.   Grass-wrack.)
    Z. marina, L.

Ruppia, L.   (Ditch-grass.)
    R. maritima, L.   Shoal water along the coast.

Potamogeton, Tourn.   (Pondweed.)
    P. natans, L.   Ditches and slow streams.
    P. Oakesianus, Robbins.   Ponds and ditches.
    P. hybridus, Mx.   (Bicknell.)
    P. amplifolius, Tuckerman.   Slow rivers.
    P. perfoliatus, L.   (Bicknell.)

Order 111.—ALISMACEÆ.   (Water Plantain Family.)

Alisma, L.   (Water Plaintain.
    A. Plantago, L.   Shallow water and edges of streams.

Sagittaria, L.   (Arrow Head.)
    S. variabilis, Eng.   Wet places.
      Var. latifolia, Gray.   With the last.
      Var. diversifolia, Gray.   With the last,
      Var. augustifolia, Gray.   With the last.
    S. calycina, Eng.   Water in flooded places.
      Var. spongiosa, Gray.   Along brooks with the last.
    S, heterophylla, Pursh.   Peekskill.   (Le Roy.)
    S. pusilla, Nutt.   Peekskill.   (Le Roy.)

Order 112.—HYDROCHARIDACEÆ.   (Frog's-bitt Family.)

Anacharis, Richard.   (Water-weed.)
    A. Canadensis, Planchon.   Slow streams.

Vallisneria, Micheli.   (Eel-grass.)
    V. spiralis, L.   Slow waters.   Along the coast.

Order 114.—ORCHIDACEÆ.   (Orchis Family.)

Orchis, L.   (Orchis.)
    O. spectabilis, L.   Shady woods.

Habenaria, Willd,, R. Br. (Rein-Orchis.)
   H. tridentata, Hook. (Leggett.)
   H. virescens, Spreng. Peekskill.
   H. viridis, R. Br. (Leggett.)
   H. Hookeri, Torr. Borders of woods.
   H. orbiculata, Torr. Pine and Hemlock woods.
   H. ciliaris, R Br. Wet places ; not common.
   H. lacera, R. Br. Damp thickets.
   H. psycodes, Gray. Wet grounds.

Goodyera, R. Br. (Rattlesnake-Plantain.)
   G. pubescens, R. Br. Shady woods.

Spiranthes, Richard. (Ladies' Tresses.)
   S. latifolia Torr. White Plains.
   S. cernua, Richard. Damp roadsides.
   S. graminea, Lindl. (Bicknell.)
      Var. Walteri, Gray. North of Kings Bridge, (Bicknell.)
   S. gracilis, Bigelow. White Plains.
   S. simplex, Gray. Woodlawn Cemetery. (Bicknell.)

Listera, R. Br. (Twayblade.)
   L. convallarioides, Hook. Damp woods.

Arethusa, Gronov. (Arethusa.)
   A. bulbosa, L. Bogs.

Pogonia, Juss.
   P. ophioglossoides, Nutt. White Plains
   P. verticilata, R. Br. White Plains.

Calopogon, R. Br.
   C. pulchellus, R. Br. Bogs.

Liparis, Richard.
   L. liliifolia, Richard. Bogs.
   L. Lœselii, Richard. (Leggett.)

Corallorhiza, Haller. (Coral-root.)
   C. multiflora, Nutt- Woods.

Aplectrum, Nutt. (Adam and Eve.)
   A. hyemale, Nutt. Rich wood.

Cypripedium, L. (Moccason-Flower.)
   C. parviflorum, Salisb. Damp woods.
   C. pubescens, Willd. Boggy places.
   C. spectabile, Swartz. Bogs.
   C. acaule, Ait. Dry woods.

Order 115.—AMARYLLIDACEÆ.—(Amaryllis Family.)

Hypoxys, L. (Star-grass.)
   H. erecta, L. Meadows and woods.

Order 116.—HÆMODORACEÆ. (Bloodwort Family.)

Aletris, L.
   A. farinosa, L. (Bicknell.)

Order 118.—IRIDACEÆ. (Iris Family.)

Iris, L. (Flower-de-Luce.)
   I. versicolor, L. (Blue Flag.) Wet places.

Pardanthus, Ker.
   P. Chinensis, Ker. Escaped from cultivation. (Bicknell.)

Sisyrinchium, L. (Blue-eyed Grass.)
   S. Bermudiana, L. Among grass.

Order 119.—DIOSCOREACEÆ. (Yam Family.)

Dioscorea, Plumier. (Yam.)
   D. villosa, L. Thickets.

Order 120.—SMILACEÆ. (Smilax Family.)

Smilax, Tourn. (Greenbriar. Cat-briar.)
   S. rotundifolia. L. Damp thickets.
   Var. quadrangularis, Gray. With the last.
   S. glauca, Walt. Dry thickets.
   S. herbacea, L. Damp meadow edges.

Order 121.—LILIACEÆ. Lily Family.

Trillium, L. (Three-leaved Nightshade.)
   T. erectum, L. Woods.
   T. cernuum, L. Moist woods.
   T. erythrocarpum, Mx. North Salem. (Mead.)

Medeola, Gronov. (Indian Cucumber-root.)
   M. Virginica, L. Rich damp woods.

Melanthium, L.
   M. Virginicum, L. Meadows, and uplands also. **Near White Plains.**

Veratrum, Tourn. (False Hellebore.)
   V. viride, Ait. Edges of meadows.

Chamælirium, Willd. (Devil's-Bit.)
  C. luteum. Low grounds ; Scarsdale. (Miss McCabe.)

Uvularia, L. (Bellwort.)
  U. perfoliata, L. Not rare.
  U. sessilifolia, L. Edges of woods and along fences.
  U. puberula, Mx. Peekskill. (Le Roy.)

Streptopus, Mx. (Twisted-stalk.)
  S. amplexifolius, DC. Damp woods.

Smilacina, Desf. (False Solomon's Seal.)
  S. racemosa. Desf Copses and along fences.
  S. stellata, Desf. Along fences,
  S. trifolia, Desf. Damp places.
  S. bifolia. Ker. Moist woods.

Polygonatum, Tourn. (Solomon's Seal.)
  P. biflorum, Ell. Edges of woods.
  P. giganteum, Dietrich. (Bicknell.)
  P. latifolium, Desf.

Asparagus, L.
  A. officinalis, L. Along the coast. (Adv. Eu.)

Lilium, L. (Lily.)
  L. Philadelphicum, L. Fields.
  L. Canadense, L. Meadows.
  L. superbum, L. Low grounds.

Erythronium, L. (Dog's-tooth Violet.)
  E. Americanum, Smith. Damp Copses.

Ornithogalum, Tourn
  O. umbellatum, L. Leggett. (Nat. Eu.)

Allium, L. (Onion Garlic.)
  A. tricoccum, Ait. Woods.
  A. vineale, L. Moist meadows and pastures. (Nat. Eu.)
  A. Canadense, Kalm. Damp meadows and rich pasture grounds.

Hemerocallis, L. (Day-Lily.)
  H. fulva, L. Roadsides. (Adv. Eu.)

Order 122.—JUNCACEÆ. (Rush Family.)

Luzula, DC. (Wood-Rush.)
  L. pilosa, Willd. Woods.
  L. campestris, DC. Fields and edges of woods.

Juncus, L. (Bog-Rush.)
  J. effusus, L. (Soft-Rush.) Wet places ; common.
  J. Balticus, Dethard. Along the coast of the Sound.
  J. Rœmerianus, Scheele. Coast of the Sound.
  J. marginatus, Rostkovius. Coast of the Sound.
  J. bufonius, L. Damp grounds.
  J. Gerardi, Loisel. (Bicknell.)
  J. tenuis, Willd. Damp fields and road-sides.
  J. pelocarpus, E. Myer. Along the coast of the Sound.
  J. militaris, Bigel. Bogs.
  J. acuminatus, Mx. Wet sandy ground.
  J. Canadensis, J. Gay. (Leggett.)

Order 123.—PONTEDERIACEÆ. (Pickerel-weed Family.)

Pontederia, L. (Pickerel-weed.)
  P. cordata, L. Edges of muddy pools.
    Var. angustifolia, Gray. (Mead.)

Order 127.—CYPERACEÆ. (Sedge Family.)

Cyperus, L.
  C. diandrus, Torr. Damp places.
  C. inflexus, Muhl. Sandy shores.
  C. dentatus, Torr. Sandy wet ground.
  C. strigosus, L. Fertile soil.
  C. Michauxianus, Schultes. (Leggett.)

Dulichium, Richard.
  D. spathaceum, Pers. Near ponds and sluggish streams.

Fuirena, Rottböll. (Umbrella-grass.)
  F. squarrosa, Mx. Wet sand.

Eleocharis, R. Br. (Spike-rush.)
  E. equisetoides, Torr. In shallow pools and flooded places.
  E. quadrangulata, R. Br. East Long Pond. (Mead.)
  E. tuberculosa, R. Br. Coast of the Sound.
  E. obtusa. Schultes. Muddy edges of slow streams.
  E. olivacea, Torr. (Leggett.)
  E. palustris, R. Br. In flooded places.
  E. tenuis, Schultes. Wet grounds.
  E. acicularis, R. Br. Edges of ponds and slow streams.

Scirpus, L. (Bulrush.)
  S. pungens, Vahl. Along the salt meadows, in the edges of pools.
  S. Olneyi, Gray. Salt marshes, coast of the Sound.
  S. Torreyi, Olney. Along the coast of the Sound.
  S. validus, Vahl. Flooded places.

S. debilis, Pursh. Swampy places.
S. maritimus, L. Borders of salt marshes along the Sound.
S. fluviatilis, Gray. Edges of Rye Lake.
S. atrovirens, Muhl. Wet swampy places.

Eriophorum, L.
E. Virginicum, L. Bogs, and wet grounds.

Fimbristylis, Vahl.
F. spadicea, Vahl. coast of the Sound.
F. autumnalis, Roem. Schultz. Wet grounds.

Rhynchospora, Vahl. (Beak-Rush.)
R. alba, Vahl, Bogs. Wet grounds.
R. glomerata, Vahl. Wet undrained lands.

Cladium, P. Browne. (Twig-Rush.)
C. mariscoides, Torr.

Scleria, L, (Nut-Rush.)
S. triglomerata, Mx. Swampy lands; common.
S. reticularis, Mx. Edges of ponds and slow streams.

Carex, L. (Sedge.)
C. polytrichoides, Muhl. Bogs. Common.
C. bromoides, Schk. Wet grounds.
C. vulpinoidea, Mx. Damp meadows.
C. alopecoidea, Tuckerman. Damp woods.
C. sparganioides, Muhl. Damp lands.
C. cephaloidea, Dew. Dry woods and fields.
C. Muhlenbergii, Schk. Dry fields.
C. rosea, Schk. Damp woods.
C. retroflexa, Muhl. Moist grounds.
C. tenella, Schk. Swamps and wet places.
C. canescens, L. Wet grounds.
C. stellulata, L.
Var. scirpoides, Gray. Wet grounds.
C. scoparia, Schk. Damp meadows.
C. lagopodioides, Schk. Wet shady places.
C. cristata, Schk. Wet grounds.
C. fœnea, Willd.
Var. sabulonum, Gray. Coast.
C. straminea, Schk. Edges of woods.
C. stricta, Lam. Wet places.
C. crinita, Lam. Along streams.
C. granularis, Muhl. Wet places.
C. pallescens, L. Meadows.
C. conoidea, Schk. Moist meadows.
C. grisea, Willd. Moist ground.
C. gracillima, Schk. Wet grounds.

C. virescens, Muhl. Dry woods.
C. triceps, Mx. Open woods.
C. plantaginea, Lam. Shady woods.
C. platyphylla, Carey. Damp woods.
C. retrocurva, Dew. Dry woods and thickets.
C. digitalis, Willd. Edges of open woods.
C. laxiflora, Lam. Woods.
C. oligocarpa, Schk. Woods.
C. pedunculata, Muhl. Woods, fields.
C. Emmonsii, Dew. Open woods and copses.
C. Pennsylvanica, Lam. Hill-sides and open woods.
C. varia, Muhl. Copses and open woods.
C. pubescens, Muhl. Damp open wood-lands.
C. miliacea, Muhl. Wet grounds.
C. debilis, Mx. Moist lands,
C. lanuginosa, Mx. Wet grounds.
C. Pseudo-Cyperus, L. Borders of ponds.
C. hystricina, Willd. Wet grass land.
C. tentaculata, Muhl.
C. intumescens. Rudge. Wet grass lands.
C. lupulina, Muhl. Wet grounds.
F. Squarrosa, L. Undrained lands.

Order 128.—GRAMINEÆ. (Grass Family.)

Leersia, Solander. (White Grass.)
   L. Virginica, Willd. Damp Woods.
   L. oryzoides, Mx. Wet grounds.

Zizania, Gronov. (Indian Rice.)
   Z. aquatica, L. Borders of sluggish streams.

Phleum, L. (Timothy.)
   P. pratense, L. Pastures and mowing grounds. (Nat. Eu.)

Vilfa, Adans. (Rush Grass.)
   V. aspera, Beauv.
   V. vaginæflora, Torr. Old sandy fields,

Sporobolus, R. Br. (Drop-seed Grass.)
   S. serotinus, Gray. Wet sandy land.

Agrostis, L. (Bent-Grass.)
   A. scabra, Willd. Dry places.
   A. vulgaris, With. (Red-top.) Grown for hay and pasture in damp lands.
   A. alba, L. (White Bent-grass.) Used for hay as the above.

Cinna, L. (Wood Reed-grass.)
   C. arundinacea, L.
     Var. pendula, Gray. Shady woods.

Muhlenbergia, Schreber. (Drop-seed Grass.)
  M. glomerata, Trin. Boggy places.
  M. Mexicana, Trin. Damp undrained grounds.
  M. sylvatica, Torr. & Gray. Wet woods.
  M. Willdenovii, Trin. Shady rocks.
  M. diffusa, Schreber. Dry woods.

Brachyelytrum, Beauv.
  B. aristatum, Beauv.

Calamagrostis, Adans. (Reed Bent-grass.'
  C. Canadensis, Beauv. Damp lands.
  C. arenaria, Toth. Coast of the Sound.

Stipa, L. (Feather-grass.)
  S. avenacea, L. (Oat-grass.) Sandy woods.

Aristida, L. (Triple-awned Grass.)
  A. dichotoma, Mx. (Poverty Grass.) Barren fields.
  A. gracilis, Ell. Near the coast of the Sound.

Spartina, Schreber. (Marsh Grass.)
  S. polystachya, Muhl. (Leggett.)
  S. juncea, Willd. Beach of the Sound.
  S. stricta, Roth.
    Var. glabra, Gray. Along the Sound.
    Var. alterniflora. Shore of the Sound.

Cynodon, Richard. (Scutch-grass.)
  C. Dactylon, Pers. (Nat. Eu.)

Eleusine, Gærtn. (Yard-Grass.)
  E. Indica, Gærtn. Door-yards.

Dactylis, L. (Orchard-Grass.)
  D. glomerata, L. Grown for hay. (Nat. Eu.)

Kœleria, Pers.
  K. cristata, Pers. Dry upland

Eatonia, Raf.
  E. Pennsylvanica, Gray. Damp woods and meadows.

Glyceria, R. Br., Trin. (Manna-Grass.)
  G. Canadensis, Trin. Wet grounds.
  G. obtusa, Trin. Swampy places.
  G. nervata, Trin. Damp grounds.
  G. aquatica, Smith. Wet grounds.
  G. fluitans, R. Br. Flooded grounds.
  G. distans, Wahl. (Leggett.)

Brizopyrum, Link. (Spike-grass.)
  B. spicatum, Hook. Shore of the Sound.

Poa, L. (Meadow-Grass.)
  P. annua, L. Cultivated grounds.
  P. compressa, L. Dry, poor land.
  P. serotina, Ehrhart. (Fowl Meadow-Grass.)
  P. pratensis, L. This is the grass used in Bourbon County, Kentucky, for pasture, and called Blue-grass; it is misnamed, for it is green. It grows well in lime-stone regions, and makes excellent pasture.
  P. trivialis, L. Damp meadows. (Nat. Eu.)

Eragrostis, Beauv.
  E. poæoides, Beauv. Sandy grounds. (Nat. Eu.)
  E. pilosa, Beauv. Sterile grounds. (Nat. Eu.)
  E. capillaris, Nees. Dry soil. Common.
  E. pectinacea, Gray. Sandy land near the Sound.

Festuca, L.
  F. tenella, Willd. (Leggett.)
  F. elatior, L. (Nat. Eu.)
    Var. pratensis, Gray.
  F. nutans, Willd.

Bromus, L. (Broom-Grass.)
  B. secalinus, L. (Cheat. or Chess-Grass.) Troublesome weed in grain fields, (**Adv.** Eu.)
  B. ciliatus, L. Damp wood-lands.
  B. sterilis, L. (Leggett.) (Nat. Eu.)

Phragmites, Trin. (Reed.)
  P. communis, Trin. Borders of ponds.

Lolium, L. (Darnel.)
  L. perenne, L. Used for pasture. (Nat. Eu.)

Triticum, L.
  T. repens, L. (Quick-Grass.) A troublesome weed in gardens.

Hordeum, L. (Barley.)
  H. jubatum, L. Shore of the Sound.

Elymus, L. (Wild Rye.)
  E. Virginicus, L. Banks of streams.
  E. Canadensis, L. Banks of streams.
  E. Hystrix, (Bottle-brush Grass. **Hedge-Hog Grass.)**

Danthonia, DC. (Wild-oat Grass.)
  D. spicata, Beauv. Among rocks.

Aira, L. (Hair-Grass.)
  A. flexuosa, L. Dry soil.

Holcus, L. (Meadow Soft-Grass.)
    H. lanatus, L. (Velvet-Grass.)

Hierochloa, Gmelin. (Holy-Grass.)
    H. borealis, Roem. & Schultes. (Leggett.)

Anthoxanthum, L. (Sweet Vernal-Grass.)
    A. odoratum. Pastures and lawns. (Nat. **Eu.)**

Phalaris, L. (Canary-Grass.)
    P. arundininacea, L. Wet grounds.

Milium. (Millet-Grass.)
    M. effusum, L. Damp woods.

Paspalum, L.
    P. setaceum, Mx. Sandy fields.
    P. læve, Mx. Frequent.

Panicum, L. (Panic-Grass.)
    P. filiforme, L.
    P. sanguinale, L. (Crab-Grass.) Weed in cultivated grounds. **(Nat. Eu.)**
    P. agrostoides, Spreng. Frequent. Wet grounds.
    P. proliferum, Lam.
    P. capillare L. Cultivated fields; common.
    P. virgatum, L.
    P. latifolium, L. Damp thickets.
    P. clandestinum, L. Damp thickets.
    P. pauciflorum, Ell.
    P. dichotemum, L. Damp grounds.
    P. depauperatum, Muhl. Hills; common.
    P. Crus-galli, L. (Barn-yard Grass.) (Nat. Eu.)

Setaria, Beauv. (Fox-tail Grass.)
    S. virticillata, Beauv. About dwellings. (Adv. **Eu.)**
    S. glauca, Beauv. In grain stubble. (Adv. Eu.)
    S. viridis, Beauv. Cultivated grounds. (Adv. **Eu.)**

Cenchrus, L. (Bur-Grass.)
    C. tribuloides, L. Near the coast of the Sound.

Andropogon, L. (Beard-Grass.)
    A. furcatus, Muhl.
    A. scoparius, Mx. (Leggett.)
    A. Virginicus, L. Sandy soil.

Sorghum, Pers. (Broom-Corn.)
    S. nutans, Gray. (Leggett.)

SERIES II.

CRYPTOGAMOUS OR FLOWERLESS PLANTS.

CLASS III.—ACROGENS.

Order 129.—EQUISELACEÆ. (Horsetail Family.)

Equisetum, (Scouring Rush.)
E. arvense, J.. Wet banks. White Plains
E. limosum, L.. West Farms. (Wood.)
E. hyemale, L. Wet banks ; common.

Order 130.—FILICES. (Ferns.)

Polypodium, L..
P. vulgare, L. Shaded rocks.

Adiantum, L. (Maiden,s hair.)
A. pedatum, L. Shady woods.

Pteris, L. (Brake.)
P. aquilina. Thickets ; common.

Cheilanthes, Swartz. (Lip-Fern.)
C. vestita, Swartz. Among rocks.

Woodwardia, Smith. (Chain-Fern.)
W. Virginica, Smith. Swamps.

Asplenium, I.. (Spleenwort.)
A. Trichomanes, L. Shaded rocks.
A. ebeneum, Ait. Woods.
A. thelypteroides, Mx. Damp woods.
A. Filix-fœmina, Bernh. Damp woods.

Camptosorus, Link. (Walking-Fern.)
C. rhizophyllus, Link. On gneisic rocks ; sparingly about White Plains.

Phegopteris, Fee. (Beech-Fern.)
P. hexagonoptera, Fee. Borders of Woods.

Aspidium, Swartz. (Wood-Fern.)
A. Thelypteris, Swartz. Marshy places.

A. Noveboracense, Swartz.   Swamps and damp woods.
A. spinulosum, Swartz.
　Var. intermedium, Gray.   About White Plains.
　Var. dilatatum, Gray.   About White Plains.
A. cristatum, Swartz.   Wet thickets.
　Var. Clintonianum, Gray.   About White Plains.
A. Goldianum, Hook.   Moist woods.
A. marginale, Swartz.   Shaded rocks.
A. acrostichoides, Swartfl.   Near White Plains.
　Var. incisum, Gray.   White Plains.

Cystopteris, Bernhardi.   (Bladder-Fern.)
　C. fragilis, Bernh.   Shaded Rocks.

Struthiopteris, Willd   (Ostrich Fern.)
　S. Germanica, Willd.   White Plains.

Onoclea, L.   (Sensitive-Fern.)
　O. sensibilis, L.   White Plains.
　Var. obtusilobata, Torr.   White Plains.

Woodsia, R. Brown.
　W. obtusa, Torr.   Cliffs and banks.
　W. Ilvensis, R. Brown.   (Leggett.)

Dicksonia, L'Her.
　D. punctilobula, Kunze.   Damp shady places.

Osmunda, L.   (Flowering Fern.)
　O. regalis, L.   Wet woods and thickets.
　O. Claytoniana, L.   Damp grounds.
　O. cinnamomea, L.   Swamps and low thickets.

Botrychium, Swartz.   (Moonwort.)
　B. lanceolatum, Angstrœm.
　B. Virginicum, Swartz.   Throughout.
　B. lunarioides, Swartz.
　Var. obliquum. Gray.   White Plains.

Order 131.—LYCOPODIACEÆ.   (Club-Moss Family.)

Lycopodium, L.   (Club-Moss.)
　L. lucidulum, Mx.   Damp woods.
　L. dendroideum, Mx.   Not common
　L. clavatum, L.   Woods.
　L. complanatum, L.   White Plains.

Selaginella, Beauv.
　S. rupestris, Spring.   Dry rocks.
　S. apus, Spring.   Damp shady places.

ADENDA.

ADDITIONS TO CATALOGUE OF PLANTS.

Lechea, L. (p. 779.)
L. Nova-Caesarea, Aust. (Leggett.)

Genista. (p. 784.)
G. tinctoria. Peekskill. State. Flora and Mead. (Adv. Eu.)

Tephrosia, Pers. (p. 784.)
T. Virginiana, Pers. Throgg's Neck. (Hall.)

Habenaria. (p. 815.)
H. viridis, R. Br.
Var. bracteata, Reich. (Leggett.)

Schollera, schreber. (p. 818, after Pontederia.)
S. graminea, Willd. Croton River. (Mead.)

Order 126.—ERIOCAULONACEÆ. (Pipewort Family.) (p. 818)
Eriocaulon, L. (Pipewort.)
E. Septangulare. Withering. Ponds—common.

Carex. (p. 819.)
C. stipata, Muhl.

ERRATA.

Cover and page 772, read Plantarum for Plantanum.

Page 774, read Miss P. A. Mc.Cabe for Miss P. A. Mecabe.

Page 775, read White Plains for Wuite Plains.

Page 776, read Marsh Marigold for Marsh Marigola, and Tulipifera for tulipifera.

Page 777, read Sarraceniaceæ for Sarracenincew.

Page 778, read Sisymbrium for Sysymbrium; and Tourn for Tonm.

Page 779, read Violaceæ for Violacea·.

Page 781, read Musk-Mallow for Mush-Mallow.

Page 782, read Tiliaceæ for Tiliacee·; and Geranium Family for Geraneum Family: and Rutaceæ for Rutacee·; and Toxicodendron for toxicodendron; and Vitaceæ for Vitaceæ.

Page 783, read Bladder-nut for Bladde-nut.

Page 784, read Lespedeza for Lespedciza.

Page 787, read Crassulaceæ· for Crasulaceæ; and Gronov for Gronor; and Live forever for live forever, and Liquidambar for Liquidamber.

Page 789, read Cicuta for Cuscuta.

Page 791, read Moist Woods for Moss Woods; and Composite Family for Composita Family.

Page 793, read X. spinosum for Y. spinosum.

Page 794, read Miss P. A. Mc.Cabe for Miss A. McCabe; and Pav. for Pax.

Page 795, read Torr. for Tow.: and Gærtn. for Gærtin; and plantaginifolia for platiginifolia; and Senecio for Senccco.

Page 796, read C. aparinoides for C. aparanoides.

Page 797, read Leucothoë for Leucothoe; and Azalea for Azalia.

Page 798, read verticillata for verticillate.

Page 799, read Raf. for Rof.

Page 800, read Toad Flax for Toad Flex; and Monkey for Monky.

Page 803, read Viper's Bugloss for Vipeis Bugloss.

Page 804, read Centaury for Centuary.

Page 811, read in Fagus ferruginea, ornamental for ornamentable: and Beech for Beach; and in Myrica cerifera, sometimes for sometime; and pulverized for puluerized; and catarrh for Cattarrah.

Page 812, read Silky for Silkey; and Aspen for Aspan; and in Pinus mitis, read Yellow Pine for Yelow Pine; an l genus for Genus; and hedging for Hedging.

Page 814, read Alismaceæ for Alismaceæ; and Plantain for Plaintain; and angustifolia for augustifolia; and Frog's-bit for Frog's-bitt.

Page 815, read White Plains for White Plalns; and verticillata for verticilata; and Moccasin-Flower for Moccason-Flower.

Page 823, read arundinacea for arundininacea; and dichotomum for dichotemum; and verticillata for virticillata.

Page 824, read Equisetaceæ for Equiselaceæ; and gneissic for gneisic.

Page 825, read Swartz for SwartH.

Page 826, read Addenda for Adenda; and bracteata tor bracleata; and septangulare for Septangulare.

NOTE.--The Report of the Flora was prepared for publication in Bolton's Revised History of Westchester County. The History was so hurried through the press that time was not afforded for second and third proofs; hence the errors now to be corrected.

--

ADDITIONS.

To Page 775, Ranunculus, L.

        R. multifidus, Pursh. (Yellow-water Crowfoot.) Scarsdale.
        (Miss Mc.Cabe.)

To Page 788, Order 40.--MELASTOMACEÆ

Rhexia, L. Meadow Beauty. Deer Grass.
R. Virginica, L. Near Rye and coast of the Sound.

To Page 788, new localities for Opuntia--Village of White Plains and North Street, three miles east of White Plains.

To Page 792, Aster, L.

        A. amethystinus, Nutt. North of Woodlawn Cemetery. (E. C. How, T. C. Bulletin.)

To Page 813, Juniperus, L.

        J. communis, L (Common Juniper.)
        Var. Alpina, L. Scarsdale. (Miss McCabe.)

☞ The author will be thankful to have his attention drawn to omissions and discoveries.